Chavan vs Hawking

Sandeep Chavan

Published by SANDEEP CHAVAN, 2025.

CHAVAN VS HAWKING

First edition. June 23, 2025.

ISBN: 979-8230947875

Written by Sandeep Chavan.

Table of Contents

To **Stephen Hawking** —

Not for what you concluded,But for what you dared to question.

Your journey redefined limits, not just of physics,but of perception, perseverance, and human will.

Though our models differ,your work lit the ripple that brought us here.

This book does not challenge your legacy.It continues the inquiry you began.

With deepest respect,—Chavan

"The edge of the universe is not where space ends—but where perception pauses."

—Chavan

Disclaimer

This book presents an alternative scientific and philosophical framework—**Universal Energy Dynamics (UED)**—developed through independent research, thought experiments, and logical reinterpretation of established theories. It is not intended to replace standard academic physics or cosmology but to **provoke inquiry, challenge assumptions, and invite deeper reflection on the nature of existence, perception, and intelligence**.

References to public figures such as Stephen Hawking are made with **great respect** and are intended to explore and reinterpret ideas—not criticize individuals. Any critique of mainstream science or theology is presented **structurally, not personally**, and stems from the author's vision of evolving thought, not confrontation.

This work is a **philosophical and scientific exploration**. It does not provide medical, technical, or legal advice. The author and publisher disclaim liability for any outcomes resulting from interpretations or applications of the concepts herein.

Readers are encouraged to engage with the content critically, cross-reference with established literature, and draw their own conclusions. The ideas presented are part of an **emerging and evolving worldview**, not a finalized doctrine.

Foreword – The Limits of Knowledge

"*To admit we do not know is not weakness. It is the only doorway through which truth ever enters.*"

There are few names in modern science that ignite as much awe and admiration as Stephen Hawking. A symbol of resilience, intellect, and cosmic curiosity, Hawking dared to ask questions that most would never dream of. He faced the void—not just the black holes he theorized, but the deeper unknowns that haunt the edges of all understanding. And with his wit, brilliance, and equations, he gave the world a bold and elegant view of reality.

But like all towering figures in science, Hawking's work—no matter how profound—is not immune to evolution. Every theory, however mathematically precise or philosophically compelling, exists within the limits of human perception, tools, and imagination at the time of its birth.

The limits of knowledge are not insults to a scientist's legacy. They are the fertile boundaries where the next question begins. Where Hawking described singularities, we now ask whether singularity itself is a misinterpretation of consequence. Where he explored event horizons, we question whether these are not edges of space, but edges of alignment—thresholds where observation itself breaks down. Where he modeled entropy and time, we now propose that entropy is not disorder, but ripple dissonance—and that time is not a dimension, but a perceptual consequence.

This book does not seek to discredit Hawking. It seeks to honor him the only way truth ever honors its predecessors—by continuing the search without fear. *Chavan vs Hawking* is not a war of ego. It is a conversation across eras. A ripple meeting a wave that came before it.

Hawking himself once said:

"The greatest enemy of knowledge is not ignorance, it is the illusion of knowledge."

If that is true, then this book is an ally to Hawking's own mission. Because illusion thrives where ripple truth remains unmeasured, and UED rises not to defeat the past, but to illuminate what it could not yet see.

As we stand at the precipice of a new paradigm—where Structured Vacuum Energy (SVE) replaces empty space, where consequence precedes cause, and where the observer is not optional, but central—we must carry forward the torch Hawking once held.

He brought us to the edge of the black hole. Now, it's time to step inside — not with fear, but with understanding.

— *For Stephen. And for what's still waiting beyond.*

Sandeep Chavan

Preface

By Sandeep Chavan – Why This Debate Matters Now

"*UED must respond — not to compete, but to complete.*"

This is not a book of defiance. It is a book of response. Not to challenge Stephen Hawking as a man, but to address the gaps that time, tools, and tradition left untouched in his grand vision of the cosmos.

For decades, we've revered black holes, the Big Bang, and thermodynamic arrows as the intellectual currency of modern cosmology. We memorized the equations, accepted the paradoxes, and stood in awe of singularities too dense to comprehend. Hawking became the voice of these mysteries — not because he claimed to solve them all, but because he dared to frame them where others remained silent.

Yet silence, over time, becomes dangerous. It breeds comfort with contradiction. It encourages reverence over re-evaluation. Somewhere between admiration and institutionalization, science stopped asking *why*, and started protecting *what already is.*

That is why this debate matters now.

Because truth evolves. And we cannot let outdated frameworks fossilize the frontier.

Universal Energy Dynamics (UED) did not emerge to overthrow existing theories with rebellion. It emerged from a necessity — the necessity to reintegrate the observer, to redefine force as consequence, and to reimagine reality not as a set of objects in space, but as a ripple field of relationships, constantly realigning within a structured vacuum.

Hawking's theories, as monumental as they are, treat the universe like a machine with broken parts — black holes, evaporating particles, missing information, and a time that may have started from nothing. But UED sees no broken parts — only misunderstood ripples. It invites us to look again, not at what's moving, but at what allows movement to be noticed at all. And that is the observer.

This debate is not urgent because Hawking was wrong. It is urgent because we have reached a point in science where even right answers are incomplete without context. Time, entropy, radiation, gravity — these concepts demand reinterpretation not in isolation, but in light of the field that births them.

UED responds not with certainty, but with consequence. It responds because the field cannot stay silent when misunderstanding masquerades as law.

So here we are — not at the event horizon of a dying theory, but at the alignment point of a new one. A ripple is not a rebellion. It's a response to imbalance.

Let this book be that ripple. Not against Hawking, but beside him — toward what lies beyond the edge of known.

— *Sandeep Chavan*

Author, Researcher, Educator

Founder of Universal Energy Dynamics (UED)

How to Read This Book

This is not a textbook. It is not a devotional. It is a conversation with the deepest truths you've always suspected—but never saw clearly articulated.

This book is for thinkers. For challengers. For those who want to break through the surface of accepted science and ask, "What really governs this universe?"

To get the most from this book:

- **Read slowly.** Many sections are not meant to be skimmed. They are engineered to provoke realization, not just deliver facts.
- **Pause after insights.** You'll encounter paradigm-shifting statements. Don't rush past them. Let the ripple of understanding settle in your own field.
- **Use the Appendices.** If a chapter bends your perspective too far, the appendices will ground you in comparisons, analogies, and redefinitions that help decode the new language.
- **Read in sequence**—but revisit freely. Each part builds on the last, but the structure allows for nonlinear contemplation. Reread chapters after the final reflection; their meaning will evolve.
- **Don't look for a god.** Look for intelligence. This book is not theological—but it is deeply spiritual in its structure. It honors both the scientific mind and the soulful questioner.
- **Talk to yourself while reading.** Debate, agree, resist. That's the ripple in action. The universe isn't asking you to

believe—it's asking you to align.

And above all:

Don't read to remember. Read to realize.

This book doesn't seek to change your beliefs.

It seeks to realign your perception.

Welcome to the ripple.

Part I: The Hole in the Horizon

Chapter 1: The Event Horizon Illusion

The boundary of a black hole is not an edge of space—it's an edge of perception, shaped by ripple tension, not gravitational geometry.

Standing at the Edge of Nothing

The scientist stood before the great void—not physically, but mathematically, philosophically, cosmically. His equations trembled at the boundary. His vision strained at the edge. Beyond this point, he believed, reality ended.

It was called the event horizon—a perfect metaphor for the final line, the ultimate curtain. A place where even light bowed out. Not because it was broken, but because the universe itself whispered: *"No farther."*

There, at that imaginary frontier, Stephen Hawking planted his flag—not as a conqueror, but as a messenger of mystery. He peered into this darkness and gave it structure. He warned us that black holes weren't just massive—they were final. Traps from which not even information could escape, threatening the very laws that science was built upon.

"It is said that fact is sometimes stranger than fiction," Hawking wrote, "and nowhere is that more true than in the case of black holes."

We believed him. We had to. Because we had no better story to tell.

But what if the horizon he saw wasn't the end of space... ...but the limit of perception? What if nothing was "trapped"—but rather unheard, unfelt, misaligned?

What if the event horizon isn't an event at all......but an illusion born from the failure to ripple back?

This chapter begins there—not with the hole, but with the edge around it. The event horizon may be the most famous boundary in science.

UED is here to tell you: That boundary doesn't exist.

Not in space. Only in our sensing of it.

What Hawking Saw: A Brief Recap of the Mainstream View

Before we challenge the foundation, we must first honor its architecture. To understand the illusion of the event horizon, we must briefly revisit how it became one of the most celebrated and feared concepts in modern astrophysics.

The Core Concepts: Singularity and Event Horizon

In the classical narrative—first formalized by Einstein's General Relativity and later reimagined by Stephen Hawking—black holes are the final state of gravitational collapse. According to this view:

- A singularity is the point at the heart of a black hole where matter is compressed to infinite density, and spacetime curvature becomes infinitely sharp. It is the theoretical limit where the known laws of physics cease to apply. A place where mass exists without volume, and gravity becomes unquantifiable.
- The event horizon is the invisible shell that surrounds this singularity. It is not a physical surface, but a mathematical boundary defined by escape velocity—the speed required to overcome the black hole's gravitational pull. Once inside,

not even light can get out. The event horizon acts like a one-way door: objects can enter, but never return. From the outside, the contents within become forever hidden.

Together, these two ideas—the singularity and its horizon—formed a model of black holes that captivated scientists and the public alike: exotic, extreme, and terrifyingly final.

Hawking's Genius: Extending the Model into Paradox

Stephen Hawking's most revolutionary work took this already dramatic picture and made it even more mysterious—and perhaps more unsettling.

1. **Hawking Radiation**: In the 1970s, Hawking shocked the scientific world by showing that black holes were not completely black. Through the application of quantum field theory near the event horizon, he proposed that black holes emit radiation—a slow, faint stream of thermal energy arising from quantum particle pair fluctuations at the edge. This meant that black holes could, over immense periods of time, evaporate.
2. **The Information Paradox**: Hawking's radiation, however, appeared to be purely thermal—carrying no information about what had fallen into the black hole. This introduced a deep contradiction: quantum mechanics demands that information must be preserved, but black holes seemed to destroy it. If an object's properties (mass, charge, spin) were lost forever after crossing the event horizon, the principle of unitarity—a pillar of quantum theory—would be broken. This became known as the black hole information paradox.
3. **Irreversibility of the Horizon**: Hawking framed the event horizon as the ultimate irreversible boundary. Once

crossed, there was no return. From the perspective of an external observer, time appears to stop as an object approaches the horizon, while from the falling object's perspective, it passes through effortlessly—but never sends signals back. This duality—accessible entry with no exit—solidified the idea of black holes as cosmic prisons.

The Framing: A Prison Without Bars, But With Absolute Laws

Hawking's black hole is not violent in the cinematic sense. It doesn't suck or scream. It simply waits—quiet, cold, inevitable. And once you've passed its threshold, you're written out of the story of the universe.

This framing has permeated not just scientific literature, but public imagination:

- A black hole is a cosmic dead end.
- The event horizon is a hard limit of nature.
- The singularity is the final insult to reason—a place so extreme that the universe itself goes quiet.

And so we built our understanding of black holes around this model: A gravitational monster sealed inside a one-way membrane, evaporating slowly while its secrets are swallowed forever.

But what if this model is a mathematical metaphor, not a physical reality? What if this "wall" is not a property of the universe, but a misinterpretation of perception?

This is where UED enters the debate—not to deny Hawking's brilliance, but to reframe the very nature of the boundary he defined.

We're not falling into a hole. We're encountering the limits of our ability to ripple back.

Let's now step beyond this metaphor—and reimagine what the horizon really is.

The Flawed Illusion: Event Horizon as an Invention of Geometry

Physics, in its modern form, often confuses the map with the territory. General Relativity, for all its mathematical elegance, is a map — a geometric model that describes how mass and energy appear to warp the fabric of spacetime. But it is not the fabric itself. This confusion lies at the heart of the event horizon illusion.

UED's Core Critique: Geometry Is Not Reality

In the Universal Energy Dynamics (UED) framework, reality is not governed by curves and coordinates—it is governed by consequence, alignment, and ripple behavior in a structured vacuum energy field (SVE). In this view:

- Spacetime geometry is a calculated convenience, a useful approximation that overlays our observations with measurable parameters like curvature, distance, and velocity. It works to predict phenomena like gravitational lensing and planetary orbits, but it does not explain what the universe *is*.
- The event horizon, therefore, is not a genuine boundary in physical space. It is a boundary in the mathematical model—specifically, a limit condition derived from the Schwarzschild solution to Einstein's equations. It defines a region where escape velocity exceeds the speed of light based on geometry, not ripple behavior or real-world field interactions.

In simple terms: The event horizon is not something you encounter. It's something your equations invent.

The Root Illusion: Misunderstanding Light

At the center of this illusion is a fundamental misunderstanding of light.

In the classical and relativistic models, light is treated as a massless particle or wave moving through space at a fixed speed — approximately 3×10^8 m/s. The entire structure of the event horizon rests on this: that once light cannot "outrun" the pull of gravity, it vanishes from reach. The boundary is then defined by the point of no return for light.

But UED redefines light, not as an object in motion, but as a perceptual alignment of ripples between the observer and the field. In this model:

- Light does not travel in the traditional sense.
- What we call light is the observer's successful resonance with a ripple emitted or reflected by another entity.
- The time we register light is not the time it took to "cross space," but the Alignment Threshold Time (ATT) — the minimum time required for ripple alignment to occur between source and observer.

So, when mainstream science says "light cannot escape a black hole," UED asks: Escape from what? And reach whom?

There is no "thing" called light trapped inside a black hole. There is only an observer failing to realign with what's beyond the ripple tension threshold.

A Thought Paradox: Is There Still a Horizon Without an Observer?

In the relativistic framework, the event horizon exists as an objective, observer-independent surface. But UED flips the frame:

What if the horizon isn't there at all unless someone is trying to perceive beyond it?

This leads to a profound reinterpretation:

- The event horizon is not a wall in space, but a limit in perception.
- It doesn't block information. It delays alignment beyond useful consequence.
- The problem is not the escape velocity—it's the ripple distortion and observational breakdown caused by extreme field misalignment.

The Geometry Trap: When Equations Replace Experience

Einstein's field equations offer a curved geometric model of how mass tells space how to bend and space tells mass how to move. But bending space is a metaphor—a clever, visually intuitive one, but still a metaphor. The universe does not curve in any physical sense. In UED, what appears as curvature is simply ripple tension realigning to restore equilibrium.

So when Hawking built the concept of the event horizon, he was not describing a wall—he was describing the limit of his model's ability to return information. The failure was not in the field, but in the geometry-based language used to describe it.

Conclusion: Geometry Explains the Shadow, But Not the Ripple

The event horizon is a shadow cast by the assumptions of geometry. It is not the end of space, not a membrane, not a trapdoor. It is the point where the ripple field's tension exceeds the observer's ability to stay in tune.

What we call the "horizon" is simply the place where resonance fails—not because the universe forbids it, but because we misunderstood what light really is.

In Hawking's world, the horizon is a black wall. In UED, it's the echo that never found its way back.

Introducing the ATT Threshold (Alignment Threshold Time)

To dismantle the illusion of the event horizon, we must first redefine what it means to *observe* anything at all. In traditional physics, observation is passive — light arrives, and you detect it. But in Universal Energy Dynamics (UED), observation is an active consequence of ripple alignment between the observer and the field. You don't "see" because something came to you. You "see" because your internal ripple structure resonated with the external disturbance.

This brings us to a core UED concept: ATT — Alignment Threshold Time.

What Is ATT?

ATT is the minimum time required for a ripple emitted (or disturbed) in the structured vacuum energy (SVE) field to realign with an observer's field of perception.

It is not absolute. It varies based on:

- Ripple density and tension in the local field
- Observer's position and internal ripple state
- Intervening distortions (field fatigue, interference, misalignment)

In simple terms:

If you perceive something, it means the ripple from that event has successfully aligned with you—within your ATT. If you don't, it doesn't mean it's not there. It means the ATT is too long, or alignment failed altogether.

The Physics Behind Perception

In the classical view, time is the independent variable. In UED, time is born from successful ripple realignment. Nothing is "happening" until something can be realigned and perceived. This is not philosophical metaphor—it is field mechanics.

Let's clarify:

- If a ripple from an event reaches you and realigns in 80 milliseconds, that becomes your experienced time delay—your ATT for that ripple.
- The ripple didn't *travel* through space; rather, you aligned with a ripple already waiting to be sensed.

This is a perception-first universe. And ATT is the pulse of your consciousness within it.

What Happens Near a Black Hole?

Now imagine a region of extreme ripple tension — a black hole.

Here, the ripple field is so dense, distorted, and compressed that:

- Ripple propagation is no longer free-flowing.
- Realignment becomes increasingly difficult.
- The ATT begins to rise exponentially.

Eventually, it reaches a point where ATT tends toward functional infinity. That is:

- It still exists.
- It's still rippling.
- But it takes so long for the alignment to happen that to any external observer, it appears as if the ripple never returns.

This is why nothing appears to escape a black hole. It's not trapped by force. It's lost to perceptual delay.

Redefining the Event Horizon with ATT

The traditional event horizon says:

"Beyond this point, escape velocity > speed of light. Nothing can return."

UED reframes it:

"Beyond this point, ATT exceeds the perceptual threshold of the observer. Realignment doesn't happen in a meaningful timeframe."

Therefore:

- Event horizon ≠ physical boundary
- Event horizon = functional misalignment threshold
- It is not a wall in space, but a drop-off in ripple compatibility
- It is not imposed by the black hole, but emerges from the interaction between the field and the observer

Implications of ATT for All Observation

This doesn't just apply to black holes.

It applies everywhere:

- You don't perceive what's too fast, too far, or too subtle—not because it's absent, but because your ATT didn't meet the requirement.
- Time itself is not a universal metronome. It is observer-specific, ripple-dependent, and field-relative.

Thus, we must abandon the illusion that perception is passive and that boundaries are static. The limits of observation are not built into the universe—they are built into ATT.

Closing Reflection: The Real Threshold

Hawking's event horizon draws a line in space. UED draws it in consequence.

"You didn't fail to escape. You just failed to align."

In the next section, we will explore what this means for the true nature of black holes: not as singularities, but as ripple tension zones in an eternally active field.

The Black Hole as a Ripple Tension Zone

"Collapse is not destruction—it is ripple compression. A black hole is not a grave. It's a resonance failure."

The black hole, in its mainstream conception, is a terrifying endpoint—a collapsed star so dense that it tears the very fabric of spacetime. At its core: a singularity where physical laws dissolve, surrounded by an invisible event horizon that seals its secrets. It is often portrayed as the ultimate predator in the cosmos, devouring light, time, and information.

But UED challenges this narrative fundamentally. Not by denial, but by reinterpretation. By looking at black holes not through the lens of geometry and force, but through the lens of ripple dynamics in Structured Vacuum Energy (SVE).

Not a Collapsed Object, but a Ripple State

In Universal Energy Dynamics, matter itself is not a thing, but a ripple pattern held in local equilibrium within SVE. When a massive star "collapses," it's not compressing into a physical point of infinite density—it's transitioning into a state of maximal ripple compression.

Here's how UED reframes the process:

- A massive star, nearing the end of its life, loses its ripple coherence.
- The collapse is not a violent explosion inward—it is a consequence of ripple failure to sustain structural tension.
- The result is a zone where ripples become so dense, so tangled, that alignment becomes functionally impossible — for the star, the field around it, and for any external observer.

This zone is what we call a black hole. But it's not a "hole" at all. It's a knot in the ripple fabric.

No Singularity, No Broken Physics

UED removes the need for singularities. Why?

Because singularities are mathematical artifacts, not physical necessities. They appear only when a theory built on continuous geometry reaches a limit it cannot resolve.

But SVE is not curved space. It is a responsive field, governed by equilibrium and consequence, not coordinates and infinities.

So, in the UED view:

- There is no "point" of infinite density.
- There is no "tear" in the fabric of reality.
- There is simply a region of extreme ripple tension, where normal dynamics collapse not into destruction, but into non-alignment.

The black hole becomes a field feature, not a fatality.

Why Light "Bends" — The Ripple Distortion Effect

In General Relativity, the curvature of spacetime causes light to follow a bent path near a massive object. Light doesn't bend because it's weak—it bends because space itself is warped.

But UED proposes a different mechanism:

- Light is not a particle traveling through space.
- Light is a ripple alignment between emitter and observer.
- Near a region of extreme ripple tension (like a black hole), this alignment is distorted, delayed, or even denied.
- The perceived "bending" of light is actually a realignment attempt across a distorted ripple topology.

From a field perspective:

The ripple didn't curve. It got compressed and twisted in the local SVE conditions. The light didn't "lose energy." The observer lost resonance clarity.

This is ripple distortion—not geometric deflection.

Gravity as a Consequence, Not a Force

General Relativity says gravity is not a force, but the effect of spacetime curvature. UED goes one step further—it removes curvature itself and roots gravity in consequence.

In UED:

- Gravity is not a pull.
- It is the directional consequence of ripple misalignment seeking equilibrium.
- In a black hole, this ripple consequence accumulates without release.
- Matter falling "inward" is not pulled—it's drawn toward a ripple sink, where alignment pressure becomes too intense to escape.

Thus:

Black holes don't trap matter with force. They absorb ripple energy that fails to re-emerge, not because it is destroyed, but because its return pathway is misaligned beyond consequence.

The Field Feature, Not the Failure

So what is a black hole, really?

- It's not a collapsed object.
- It's not an infinite point.

- It's not a trapdoor to oblivion.

It is a ripple anomaly, a zone of extreme misalignment, a consequence echo of a field that was once stable but is now overwhelmed by internal tension.

It doesn't violate laws—it enforces them, just at a magnitude beyond what our current observational bandwidth (and equations) can resolve.

"A black hole is not the end. It is a ripple refusing to speak back." "Not because it cannot. But because we no longer share a language of alignment."

A New Boundary: Not of Space, But of Sensibility

What if the boundary we've feared in black holes—the infamous event horizon—was never a boundary in space at all, but a boundary in sense? Not the edge of the universe, but the edge of our ability to participate in it?

This is where Hawking's cosmic prison ends, and UED's field-conscious framework begins.

Redefining Inaccessibility: From Geometry to Resonance

In Hawking's view, the event horizon is an invisible but absolute wall:

- It defines the point of no return.
- Once crossed, not even light—or the information it carries—can escape.
- To the external universe, the interior becomes forever inaccessible.

This “inaccessibility” becomes a philosophical trauma in modern physics—where does the information go? What happens beyond the horizon? Are we doomed to never know?

But UED reframes the whole premise:

It’s not that we can’t access the interior. It’s that we can no longer align with it.

There is a profound difference.

In UED:

- There is no hidden space—only ripple regions that exceed relational thresholds.
- There is no absolute cutoff—only zones of failed resonance.
- What Hawking labeled as “inaccessible” is actually temporarily unresolvable from the observer’s field of perception.

This isn’t about walls. It’s about compatibility.

Interaction Limits, Not Spatial Boundaries

Think of the universe not as a container, but as a field of ripple interactions:

- Every observable phenomenon arises from successful ripple realignment between source and observer.
- When that alignment fails—not because of distance, but due to field tension and distortion—we don’t stop receiving; we simply stop understanding.

Thus:

- The event horizon is not a "where."
- It is a "when and how" that no longer functions between systems.
- It is not a fixed place in space; it is a limit of ripple transaction.

Like a radio that can no longer tune into a signal, the observer finds silence—not because the station stopped broadcasting, but because the field is too distorted to resonate anymore.

Sensibility Over Spatiality

This leads us to one of the most radical claims in UED:

The universe is not divided by space. It is divided by sensibility.

Sensibility here means:

- The capacity to align with ripple reality
- The readiness to receive and translate field information
- The consequence of failing to tune into a distorted zone

This makes the so-called "boundary" psychological, physical, and field-relative—not a universal truth, but a conditional threshold of interaction.

You cannot cross the horizon because you and the field are no longer in phase.

The Observer's Role Reinstated

General Relativity tends to treat the observer as a passive follower of geodesics—a body moving along the curves of space. Hawking's framework is more abstract, distancing the observer even further.

But in UED, the observer is central. Not because it wants to be, but because perception is real physics.

If a ripple doesn't align with an observer, it doesn't "exist" in any meaningful way for them. Reality is built on relational alignment, not inert objects moving through voids.

Therefore:

The horizon is not the field's barrier. It is the observer's moment of failure—to stay in tune.

The Real Horizon Is Within

We have been looking at black holes as cosmic vaults that hide their contents behind mathematical barriers. But perhaps what's hidden is not information—but our own assumptions.

"The horizon doesn't hide what's inside. It reflects our inability to align with what's within." — Sandeep Chavan

That failure is not irreversible. It's not divine. It's not final. It's relational.

And that changes everything.

Consequence Over Collapse

"Black holes don't collapse reality. They store unresolved consequence."

The prevailing image of a black hole is a collapsing nightmare — a gravitational implosion so violent and absolute that even light succumbs to its pull. But what if the collapse is not collapse at all? What if this terrifying finality is actually a state of unresolved field tension?

This is the fundamental shift UED offers: We move from a model of gravitational death to one of ripple consequence. We stop talking about "trapped energy" and start understanding held tension.

Collapse Is Not Catastrophe — It's Congestion

In General Relativity and Hawking's framework:

- A black hole forms when mass collapses under its own gravity.
- This creates a singularity, a point of infinite density and zero volume.
- All energy, mass, and information that crosses the event horizon is effectively *lost to the universe*, locked in forever.

But in UED:

- There is no singularity.
- There is no trapped mass.
- There is no forced collapse in a geometric sense.

Instead, a black hole is a consequence accumulation zone:

- It forms when ripple disturbances in SVE can no longer self-correct in real-time.
- The field does not "implode" — it enters a state of tension saturation.
- The region becomes unable to dissipate energy through standard ripple channels.

This doesn't mean collapse. It means congested feedback.

The field still holds the energy, but in a compressed, high-tension ripple configuration.

Ripple Exhaustion, Not Hawking Radiation

Hawking proposed that black holes radiate energy due to quantum fluctuations near the event horizon — a virtual particle-antiparticle pair separation, where one falls in and the other escapes, appearing as thermal radiation.

This elegant idea leads to Hawking radiation and eventual black hole evaporation. However, it creates paradoxes:

- Why is the radiation purely thermal with no information content?
- How is causality maintained when particles are separated across a boundary?
- How do we reconcile this with unitary quantum mechanics?

UED bypasses these issues with a simpler proposition:

Black holes don't radiate via particle escape. They exhaust ripple tension over time.

This exhaustion happens when:

- The field reaches localized ripple release thresholds.
- SVE slowly begins to offload stored ripple density, not as high-energy particles, but as gradual field relaxation.

This release is not thermal, not quantum, and not particle-based. It is field behavior—subtle, continuous, and distributed across ripple harmonics.

In this model:

- There's no mystery about where the energy goes.

- There's no paradox about what escapes and what doesn't.
- There's only a slow return to equilibrium, not a mechanical decay.

Irreversibility vs. Conditional Return

One of Hawking's boldest conclusions was that the event horizon is irreversible. Once information crosses it, there's no coming back—not even in principle. This drove the famous black hole information paradox and initiated decades of theoretical struggle.

UED radically reinterprets this.

We say:

Irreversibility is not a law. It's a consequence of failed alignment.

In other words:

- The threshold (mistaken for a "horizon") is not a one-way wall.
- It is a zone of misalignment that can become reversible if ripple resonance is re-established.

When the local ripple field:

- Relaxes from extreme tension,
- Realigns with nearby field domains,
- Or reenters compatible resonance with an external observer...

...the perceived "loss" ends. The information—or more accurately, the consequence—resurfaces, not through recovery, but through field reintegration.

This is not science fiction. This is ripple logic in structured vacuum dynamics.

Reality Is Not Collapsing. It's Waiting.

Black holes don't erase the universe. They hold its delayed effects. They don't trap light. They create zones where light cannot speak to us. They don't destroy information. They store it outside the current ripple bandwidth.

Hawking's horizon was a verdict. Chavan's threshold is a question: *Can you realign?*

And if the answer is yes, then what we once called irreversible is actually a ripple looping back to coherence.

Closing Reflections: From Illusion to Insight

"There is no edge in the universe—only the point where your ripple stops listening."

We've spent this chapter peeling away one of the most iconic illusions in modern astrophysics: the event horizon.

Not to discard Hawking's brilliance—his contributions opened the door—but to acknowledge that the door was never the end. It was simply our pause in understanding. A threshold mistaken for a finality.

So before we leave this chapter behind, let's ask the question again—this time with the lens of ripple dynamics, field interaction, and observer alignment fully in focus.

What If... Black Holes Aren't the End?

What if we've misread black holes not because we lacked data, but because we lacked perceptual context?

What if black holes are not singularities of destruction, but zones of potential restoration—where unresolved ripple tension stores consequence, waiting not to annihilate, but to realign?

What if the famous "event" in event horizon was never about crossing a boundary— ...but about failing to continue perceiving beyond a certain point of ripple distortion?

- It's not that things fall in and vanish.
- It's that we stopped resonating with their presence.
- The problem isn't gravity pulling.
- It's our ripple field disconnecting.

Just like grief isn't about someone disappearing—but about your inability to feel them anymore—so too is the black hole a loss of resonance, not a loss of reality.

Rethinking Finality

We assumed that the moment an object disappears beyond the event horizon, it's lost forever. But UED invites you to consider:

Maybe that moment wasn't their *end*. Maybe it was *our failure to continue the conversation.*

Just because the field goes silent doesn't mean it stopped existing. It means your ripple can't hear it anymore. But given time, given restoration of field equilibrium, the silence can be lifted.

Where Hawking drew a line in space, UED draws a curve in consequence.

From Illusion to Insight

The real illusion isn't the black hole.

It's the certainty with which we claimed to know what lies inside it—using tools that couldn't see, hear, or ripple that far.

Insight begins when we release the illusion of boundary and embrace the reality of conditional perception.

There are no hidden worlds, only worlds misaligned with our awareness.

And that should humble us—not frighten us.

Because it means the universe isn't divided into accessible vs. forbidden... ...it's divided into what we've learned to ripple with, and what we haven't yet remembered how to feel.

The Path Ahead

In the next chapter, we'll strip away the biggest myth of all: That at the heart of a black hole lies a singularity.

We'll show why singularities never existed, why they are theoretical failures rather than physical realities, and how Structured Vacuum Energy makes them obsolete.

We'll move from:

- Collapse → Consequence
- Infinity → Misalignment
- Mystery → Measurable Misperception

Because where Hawking reached a dead-end, UED finds a doorway — shaped not by gravity, but by the ripple memory of what we refused to align with.

"There is no edge in the universe—only the point where your ripple stops listening." — Sandeep Chavan

Chapter 2: No Singularity, No Surprise

The idea of a singularity arises from the failure of geometry—not from a fundamental truth of nature. UED replaces the concept of infinite curvature with ripple density and field saturation.

The Myth at the Center

Imagine falling—not toward the Earth, not into water or atmosphere, but toward something far more abstract: a place where all coordinates collapse, all equations diverge, and all known physical laws surrender.

This is the mythic center of a black hole: A point so dense, so extreme, so unrelenting in its gravity, that it defies comprehension.

In Hawking's world—and the world of General Relativity—the singularity is both mathematically demanded and physically unreachable. It is the end of calculation and the beginning of speculation. A conceptual black box where reality is said to bend infinitely into itself, leaving behind only the certainty that we no longer know.

You can't see it. You can't describe it. But the model insists it's there. It sits at the heart of the black hole like a god in exile—feared, unseen, unquestioned.

But what if it was never real to begin with?

What if the singularity is not the universe's secret chamber, but a monument to the limits of our models?

What if the infinite curvature, the zero-volume density, the breakdown of laws — were all symptoms not of nature, but of our misinterpretation of it?

This is where Universal Energy Dynamics offers its correction.

Not by imagining a new type of singularity, but by eliminating the need for one entirely. Not by patching the equations, but by changing the question.

In UED, what lies at the center of a black hole is not a singularity—but a zone of ripple saturation, where tension exceeds structure, but never escapes consequence.

This is not a mystery. It is a mechanism.

Let's now step into the heart of the illusion—and redefine the center not as a failure, but as a feature of a field still functioning exactly as it must.

Because where equations diverge, ripples compress. And where physics breaks, consequence still unfolds.

What Is a Singularity? (In Classical and Relativistic Physics)

"The singularity was never a discovery. It was a confession—written in the language of math."

The term singularity carries a certain reverence in the scientific imagination. It evokes awe, fear, mystery—a place where even the universe seems to hold its breath.

But let's step back and ask a simple question: What, exactly, is a singularity?

Defining the Singularity

In the framework of General Relativity, a singularity is the endpoint of gravitational collapse—a theoretical location in spacetime where the known laws of physics cease to function. It is defined not by what we know, but by what we can no longer compute.

More precisely:

- It is a point of zero volume yet infinite density. All the mass of a collapsing star is said to compress into a region of no spatial extent, producing gravitational forces beyond limit.
- It is a mathematical consequence of Einstein's field equations, which—under certain conditions—predict that curvature of spacetime becomes infinite at a central point within a black hole.
- It is the breakdown of the curvature model itself. When you trace the geodesics of particles or light back toward the core of a black hole, the paths end abruptly—not because there's a surface or a wall, but because the model stops making sense. You encounter an undefined point where distance, time, mass, and force no longer behave predictably.

In short, the singularity is not a feature of reality. It is a hole in the map.

The Problem: Physics Ends at the Singularity

This should deeply concern us.

Physics is a science of predictive behavior. It's meant to offer continuity, causality, and comprehension. But the singularity is none of those things.

It's not a physical place. It's not an observable object. It's not even testable.

It's a mathematical artifact—an outcome of extrapolating equations beyond their applicable domain. The moment the model reaches its structural limit, it doesn't offer a refined explanation; it throws up an infinity and walks away.

This is not a breakthrough. This is a breakdown.

And yet, we've built entire cosmologies around it. Black holes. Big Bangs. The edge of time itself—all constructed on the back of a concept that is, at its core, an admission of ignorance.

Einstein himself was cautious about this. He suspected singularities were indicators of model failure, not physical truths. Later physicists, however, took them literally, embedding them into the heart of theory and fiction alike.

The Dangerous Appeal of Infinity

Why does the singularity still persist in our theories?

Because it's seductive.

It offers a dramatic climax to the cosmic story:

- A place where laws vanish
- Where time ends
- Where the universe "began" or "collapses"

But this is narrative, not nature.

Nature doesn't do infinite curvature. It doesn't create points of zero volume with limitless mass. These are pathological results from mathematical models that have exceeded their valid boundary conditions.

Just like dividing by zero gives you a meaningless answer, so too does pushing Einstein's equations to the limit give you a physical impossibility disguised as a scientific conclusion.

Why UED Rejects the Singularity Entirely

This is where Universal Energy Dynamics draws the line.

In UED, there are no singularities, because there are no infinities in real field dynamics. What appears as a singularity in geometric terms is, in UED, simply a region of extreme ripple saturation—a field configuration where ripple density exceeds conventional thresholds, but never escapes causality.

The laws of physics don't break there. They simply change form—from structure to compression, from flow to tension.

So instead of looking at the heart of a black hole and seeing the end of everything, UED looks and sees a different kind of order—waiting to be understood.

The singularity didn't break the universe. It broke our equations.

And that's not nature's failure. That's ours.

The Failure of the Equation — Not the Universe

"Infinity is not a destination. It is a red flag."

When a theory reaches its limit, it doesn't always say so with humility. Sometimes, it shouts back absurdity.

This is precisely what happens in the case of singularities. They are not physical phenomena waiting to be discovered—they are mathematical artifacts, warning signs that we've taken our equations beyond their legitimate domain of application.

Equations Have Limits. Reality Does Not.

The field equations of General Relativity are incredibly powerful. They describe planetary motion, gravitational lensing, time dilation, and even the expansion of the cosmos. But like any tool, they are valid only within the assumptions that define them.

- They assume a smooth, continuous spacetime manifold.
- They treat mass and energy as idealized entities in a clean field.
- They ignore quantum effects at extremely small scales.

When those assumptions no longer hold—say, inside the core of a black hole—the mathematics doesn't break nature. It simply ceases to represent it.

Singularities arise not because the universe ends, but because the math runs out of breath.

In that moment, what we get is not clarity, but an infinity. Not a deeper truth, but a placeholder for ignorance.

Geometry Collapses, Not Matter

In the relativistic view, as you approach the center of a black hole, spacetime curvature increases until it becomes "infinite." This is interpreted as the formation of a singularity.

But let's ask plainly:

- Is anything actually *there* at that point?
- Is matter "infinitely dense"?

- Is time truly gone?

No. What's "there" is simply a breakdown in geometric description.

The fabric we used to measure and predict—spacetime—has stopped behaving in a way our tools can meaningfully describe. But that doesn't mean matter has ceased to be matter, or that energy has been compressed into a fantasy point. It only means our *model of geometry* no longer matches the field's behavior.

UED reframes this beautifully:

Geometry is a map, not the terrain. And sometimes, when the terrain gets too steep, the map folds in on itself.

The Analogy: Dividing by Zero

Imagine standing in front of a chalkboard and watching someone perform this equation:

$$\frac{1}{x} \quad \text{as } x \to 0$$

The answer tends toward infinity. Not because there's some deep physical truth emerging, but because you're trying to divide by something that no longer makes sense.

In mathematics, infinity in this context doesn't describe reality—it marks the edge of the equation's applicability. It signals a breakdown in representation, not a breakthrough in understanding.

The singularity, in this sense, is our chalkboard infinity. We divided by zero and called it a cosmic law.

But the universe never divides by zero. It redistributes tension, saturates fields, and transitions between ripple states—without ever losing continuity.

UED's View: Field Continuity, Not Mathematical Collapse

In Universal Energy Dynamics, ripple behavior remains fully governed by consequence, interaction, and equilibrium—even under extreme tension.

There is no need to invent a "place" where all rules stop. There is only the need to recognize when our language of description—geometry, equations, idealizations—no longer matches the reality we're trying to describe.

Singularities don't tell us what's *there*. They tell us what our current theory couldn't yet hear.

Final Thought: The Universe Didn't Fail—We Did

We should stop treating singularities as discoveries and start treating them as diagnostic errors—moments when the model didn't reflect the field, and we mistook that for nature's collapse.

"The singularity is not the universe's limit—it's ours." "And what breaks down is not matter—but the mirror we tried to trap it in."

UED Response: No Singularities, Only Ripple Saturation

"Where others see an impossible point, UED sees a saturated field—too full to collapse, too tense to echo back."

The moment modern physics says "singularity," UED says: not necessary.

In place of an undefined mathematical abyss, UED offers something far more grounded and physically consistent—Ripple Saturation.

No infinities. No zero-volume riddles. No broken rules.

Just a field pushed to its dynamic limit, doing what fields always do: Storing tension until it can redistribute.

The Ripple Saturation Zone

At the heart of a black hole—where relativity places its singularity—UED locates a Ripple Saturation Zone (RSZ).

This is a region where:

- The ripple activity within Structured Vacuum Energy (SVE) has reached its maximum compression.
- Local field tension has intensified beyond the capacity for standard structural stability.
- The result is not collapse into "nothing," but a temporary state of ripple congestion, where normal field behavior—motion, interaction, resonance—is suppressed or frozen.

It's not a point. It's a state—a ripple configuration so tightly packed that coherent structure ceases, not because matter has been destroyed, but because form is overwhelmed by frequency.

Maximum Field Compression ≠ Infinite Density

UED categorically rejects the idea of infinite density. The universe, in its most extreme states, still obeys consequence and capacity.

- Density in UED is not measured by mass per unit volume, but by ripple tension per field domain.

- The saturation point is not an asymptotic explosion toward infinity—it is a ceiling defined by the structure of SVE itself.

In short:

There's a maximum amount of tension the field can accommodate without releasing or reconfiguring. That's ripple saturation—not collapse, and certainly not a singularity.

This zone is measurable in principle, reversible under the right conditions, and always governed by conservation, not destruction.

Dissolution, Not Collapse

What happens to matter that enters such a zone?

Mainstream theory says:

- Matter collapses into a point.
- Volume disappears.
- Geometry breaks down.

UED says:

- Matter loses form.
- It doesn't collapse—it dissolves.
- It transitions from structured, stable ripple states (mass-bearing configurations) into non-coherent ripple tension, spread across the saturation zone.

Imagine dropping a complex structure into a whirlpool. It's not erased—but its pattern is overwhelmed by turbulence. That's what happens in ripple terms.

No destruction—only de-structuring.

And this ripple soup doesn't vanish. It waits—compressed, encoded, and ready for reintegration when the field allows.

From Collapse to Consequence Management

The RSZ (Ripple Saturation Zone) acts like a cosmic pressure chamber:

- It absorbs ripple load.
- It resists reorganization due to excessive internal tension.
- It becomes "invisible" to external observers—not because of gravity, but because the ripple misalignment exceeds Alignment Threshold Time (ATT).

This is where the illusion of "nothing escapes" arises.

Not from an event horizon. But from the ATT blackout caused by ripple saturation.

A Field Under Stress, Not a Lawless Singularity

UED doesn't require a singularity to explain black holes because the universe doesn't require it. The concept was only ever needed because we forced the wrong tools onto the problem.

When viewed through the lens of ripple dynamics:

- The "singularity" is a ripple zone under tension, not termination.
- The breakdown is in form, not in law.
- The core is not a point of infinite density, but a domain of field saturation, fully within the rules of energy conservation, consequence, and ripple equilibrium.

"The center of a black hole is not the death of physics. It is a consequence reservoir, waiting for balance to return."

And when that balance does return—whether in this epoch or another—the field will ripple once again.

Ripple Density vs. Curvature

"Curvature is not a thing. It's a shadow—cast by deeper ripple misalignment."

One of the grandest ideas in Einstein's General Relativity is that mass curves spacetime, and that curvature tells matter how to move. In this picture, the fabric of reality is bent and twisted around massive objects, giving rise to the phenomenon we call gravity.

This geometric interpretation was revolutionary. It replaced Newton's concept of gravity as a force with a more elegant framework: geometry itself became causal.

But there's a catch.

While the mathematics of curvature works well on large scales, it becomes ambiguous, abstract, and ultimately meaningless under extreme conditions—such as near black holes or inside the so-called singularity.

This is where Universal Energy Dynamics (UED) steps in with a deeper lens.

Ripple Density: A More Foundational Measure

In UED, we do not define reality by curved space, but by the behavior of ripples within Structured Vacuum Energy (SVE).

And to understand what replaces "curvature" in UED, we introduce a key concept: Ripple Density.

Ripple density refers to:

- The localized intensity of ripple compression within SVE.
- A direct measure of field tension, distortion, and structural misalignment.
- It does not describe an abstract bending of coordinates. It describes how stressed the field is, based on the overlapping and interacting ripple patterns.

It serves a role analogous to mass, but it's not about quantity—it's about quality of alignment:

- A region with high ripple density is under severe internal tension.
- The system struggles to maintain coherent ripple structures like matter, charge, or radiation.
- It is approaching—or has reached—saturation, where ripple interaction becomes chaotic or suppressed.

So while mass is a parameter of quantity and position in classical physics, ripple density is a parameter of consequence and internal resistance.

Why Curvature Is Just a Shadow

In Einstein's theory, gravitational effects arise from how spacetime is curved. But UED reveals this as a projection of deeper field interactions.

Curvature is not the cause—it's the symptom of ripple dynamics beneath it.

What appears as "curved spacetime" in general relativity is, in UED, the visible shadow of ripple consequences that the equations of relativity have approximated geometrically.

Let's be clear:

- Curvature describes what test particles appear to do.
- Ripple density describes why the field behaves that way in the first place.

When ripple density is high:

- The field distorts.
- Ripple pathways bend, delay, scatter, or cancel.
- Observational effects such as gravitational lensing, time dilation, and energy redshift appear—not because space is bending, but because alignment pathways are becoming more stressed.

This gives rise to an optical and mathematical illusion of curvature.

Infinite Curvature? Just Extreme Misalignment

When General Relativity reaches its mathematical edge—typically near the center of a black hole—it claims infinite curvature.

But in UED, we reinterpret this:

- What we're seeing is not a geometric crisis, but a consequence overload.
- Ripple tension has exceeded the field's ability to reorganize smoothly.
- Alignment pathways break down, not space itself.
- The spike in "curvature" is just a visible echo of extreme

ripple misalignment pressure.

Thus:

What we once called infinite curvature is simply a localized ripple condition where all incoming and outgoing alignment attempts are overwhelmed.

This is ripple logic. This is field behavior—not broken geometry.

The Observer's Perception and the Illusion of Geometry

Why does this matter?

Because observers do not see fields directly—they perceive consequence. Curvature is one way of *describing* those consequences. But ripple density is what causes them.

UED restores physics to its causal foundation:

- It recognizes that every "curved" motion is a response to ripple topology.
- It shows that matter follows not geometry, but alignment gradients.
- The observer senses time delay, energy distortion, or deflection—not because of a spacetime well, but because ripples are trying to find coherence under stress.

Final Insight: Replace Geometry with Behavior

Let's be honest: Spacetime curvature was a brilliant model. But it was never the thing itself.

Ripple density is not a metaphor. It is a real, measurable condition of the universal field.

"What bends is not space. What stretches is not time. What shifts is the ripple beneath your every perception— ...seeking balance in a field under pressure."

In the next section, we'll see what happens when that pressure peaks—not in collapse, but in transformation. Because physics doesn't break at the center of a black hole.

It redistributes.

No Physics Breaks — Only Ripple Overload

"Nature never breaks. Our equations do."

At the center of a black hole—where General Relativity halts in silence, and quantum mechanics offers only speculation—most physicists agree on one thing: we don't know what's going on.

The dominant narrative says physics itself "breaks down" at the singularity. It's spoken like a cosmic surrender, with equations yielding only infinities and models dissolving into paradoxes. The idea is that space and time stop behaving, matter loses meaning, and the very framework of cause and effect implodes.

But Universal Energy Dynamics (UED) does not accept this breakdown. Because nature doesn't break. It adapts. It transforms under pressure. It reorganizes under constraint. It never abandons consequence—only our models do when they can't keep up.

Let's reframe this once and for all.

Singularity Models Declare a Dead-End

In mainstream singularity theory:

- Curvature becomes infinite

- Density becomes undefined
- Time stops or becomes imaginary
- Predictability ends

These aren't just extreme conditions—they're declarations that our mathematical tools have hit their wall. It's not that the universe ends. It's that our representation of it ceases to apply.

And yet, instead of admitting the limitation, science has long mythologized it:

"Here lies the singularity—physics rests in peace."

But UED brings it back to life. Because the field never stopped working. It only stopped aligning with the assumptions of curved geometry and clean equations.

UED: Continuity in the Face of Compression

In UED, nothing truly collapses. It compresses. Nothing vanishes. It reorganizes. No fundamental law ends. The conditions for its manifestation shift.

Even in the most extreme ripple states—like the core of a black hole—the field:

- Still exists
- Still stores ripple
- Still obeys ripple dynamics and consequence

It's just that:

- The ripple configurations become so dense, so interlocked, that traditional physical behaviors—motion, radiation, signal exchange—are suppressed.

- Ripple alignment becomes functionally impossible for external observers. Not because there's a wall, but because the Alignment Threshold Time (ATT) exceeds perceptual capacity.
- Local ripple structures enter a state of dynamic stasis: not death, not destruction, but temporary non-expression.

Think of it as a storm of information—so intense, so tightly packed, that coherence is no longer visible, though everything still *is*.

Physics Doesn't Break — It Becomes Nonlinear

Much like sound distorts in a thickening medium, or like light slows in a dense refractive material, ripple behavior in UED responds to environmental conditions.

Under extreme saturation:

- Ripple waves bend, scatter, cancel, or entangle
- Coherent ripple loops (which we interpret as particles, waves, or structures) lose form
- Field dynamics continue, but their observable expressions disappear into nonlinear behavior

From the outside, it looks like nothing is happening. But from the field's perspective, it's a reconfiguration under stress.

This is the heart of UED's strength:

It never claims a breakdown of law—only a shift in ripple logic.

We Must Stop Blaming Nature for Our Failures

The singularity is often treated like a physical feature of the universe. But it's really a mathematical scar, left by our attempt to stretch a model past its limits.

What do we really mean when we say "physics breaks"? We mean:

- "My equation doesn't resolve."
- "My concept of space fails."
- "My curvature model gives me infinity."

But nature doesn't care. The ripples are still there. The field is still real. The universe doesn't deal in infinities. It deals in balance, saturation, and consequence.

Final Thought: Don't Rewrite the Universe—Rewrite the Equation

Singularities are not found in telescopes. They're found in textbooks. And while telescopes show silence at the center of a black hole, UED sees a field under pressure—not a failure, but a feature.

"Nothing is missing. Everything is misaligned." "The ripple still holds the message. You just stopped speaking its language."

In the next section, we explore how time itself responds to these ripple pressures—and why the "stopping of time" near a black hole isn't an act of physics breaking, but the observer losing sync with the field clock.

Time and the Observer Near the Core

"Time doesn't stop. You simply stop syncing with it."

One of the most puzzling and poetic ideas in modern physics is that time slows down near a black hole—and, at the singularity, time ends. From a classical observer's standpoint, objects appear to freeze as they approach the event horizon. From the falling object's frame, time ticks on—until it too reaches the core, where the equations of relativity suggest time itself becomes undefined.

This time dilation near black holes has long been interpreted as a warping of time by gravity, a stretching of temporal intervals due to extreme curvature.

But what if this is just an illusion created by perceptual breakdown?

What if the "slowing of time" isn't because time is bending or ending—but because ripple alignment between field and observer fails?

In Universal Energy Dynamics (UED), time is not an independent entity flowing uniformly through the cosmos. Time is an emergent consequence of ripple realignment—quantified by ATT: the Alignment Threshold Time.

When Ripple Density Rises, ATT Stretches

In UED, ATT is the minimum amount of time required for a ripple disturbance in the field to realign with the observer's perceptual frame. It defines the interval between ripple cause and ripple consequence, between signal and sensation.

As ripple density increases—especially near the core of a black hole—the field becomes:

- Hyper-compressed
- Internally entangled
- Saturated with misaligned ripple pressure

In such an environment, ripple realignment becomes exponentially harder. This causes ATT to stretch—no longer milliseconds or seconds, but hours, years, or functionally infinite.

The signal hasn't stopped. The path to receive it has collapsed under misalignment tension.

Time Slows—Not Because It's Warped, But Because You Are

For the external observer watching an object fall toward a black hole, it appears to slow down—freezing at the horizon. Why?

Because the ripples from the infalling object take longer and longer to align with the external field. The observer's ATT to the object increases—not because the object's time is ending, but because their perceptual bridge is collapsing.

From the infalling object's point of view, time appears to move normally—until, at the core, even internal ripple structures lose coherence, and perception becomes impossible.

This is not timelessness as an absolute. It's a state of unresolved alignment.

"The core doesn't stop time. It stops your ability to feel it."

The "Timeless Core" Is a Misinterpretation

In classical theory, the singularity is where time itself ceases to exist—a place where spacetime curvature becomes infinite and causality dies.

But in UED, nothing disappears—not time, not space, not cause. What disappears is our access to the field through ripple alignment.

Thus:

- The core is not timeless in nature.
- It is timeless to the observer—because ripple density has reached a point where ATT exceeds any functional threshold of experience.

This is crucial:

- Time is not gone.
- It is unresolvable at that location from your frame.

The perception of time ending is the experience of ripple desynchronization, not universal cessation.

The Singularity Is a Perceptual Horizon

This reframes the singularity in its entirety.

The "point" at the center of a black hole is not a physical pin-prick of infinite collapse. It's a perceptual dead zone—a horizon created not by geometry, but by the limits of relational sensing.

The further into ripple saturation we go:

- The more ATT stretches.
- The less coherent feedback we get.
- The more the region appears "silent," "still," and "infinite."

But the universe didn't disappear. You did—from its view, and it from yours.

"The singularity is not at the center of a black hole. It's at the edge of your capacity to relate."

Final Reflection: Alignment Is Time

Time, in UED, is not some river flowing independently. It is the rhythm of ripple consequence, timed by ATT. And when ATT becomes too large to resolve, the rhythm stops—not in the universe, but in you.

This is why UED insists: The universe never ceases. Only perception stalls, and with it, the illusion that time has died.

"You didn't cross a line where time ended. You crossed a threshold where alignment paused."

Let's now move forward to Section: The Philosophical Misdirection of the Singularity, where we examine why the very concept of the singularity became a scientific myth—and how it seduced generations into mistaking limitation for profundity.

The Philosophical Misdirection of the Singularity

"The singularity is not a revelation. It's a resignation—dressed in equations."

There are moments in science where abstraction begins to look like depth, and where mystery is mistaken for truth. The singularity is one such moment.

Originally introduced as a mathematical warning—a sign that an equation has exceeded its valid domain—the singularity has, over time, been elevated to something more: a metaphysical cliff, a sacred unknown, a sort of scientific Nirvana where the rules dissolve and possibilities explode.

This transformation didn't occur because of observation or evidence. It occurred because physics reached its conceptual edge and turned that failure into a symbol.

The Singularity as a Placeholder for the Unknown

In classical general relativity, the singularity emerges naturally when equations are pushed into extremity—particularly when modeling gravitational collapse.

- Curvature becomes infinite.
- Density becomes undefined.
- Time and space lose distinct meaning.

This was not meant to be a physical prediction. It was a mathematical caution sign. But over time, it became rebranded as a physical destination.

Scientists began saying things like:

- "This is where the universe began."
- "This is the heart of every black hole."
- "This is where time ends."

But what was really happening?

We mistook a gap in understanding for a profound truth.

The singularity was never the answer. It was the boundary of the model—the place where the map stops matching the terrain. But instead of finding better maps, we turned the edge into a mythological marker.

Mystery Without Mechanism

The singularity seduces because it is dramatic:

- A point of infinite density!
- A place where physics collapses!

- A region outside time!

But for all its poetry, it offers nothing concrete:

- No structure.
- No causality.
- No process.

You cannot explain what happens at a singularity—you can only say that explanation itself fails.

This is not science. It is philosophical retreat disguised as theoretical insight.

Worse, the concept of the singularity invites abstraction to replace investigation. When faced with breakdowns, researchers defer to quantum gravity, string theory, or speculative mathematics—without ever revisiting the root assumption that perhaps singularities aren't real at all.

UED Strips Away the Illusion

Universal Energy Dynamics (UED) refuses to romanticize failure. Where general relativity throws up an infinity, UED delivers a measurable saturation. Where classical physics invokes mysticism, UED invokes ripple logic and field mechanics.

In UED:

- There is no need for infinite curvature—just ripple tension thresholds.
- There is no point of no return—just zones of misalignment beyond observer resolution.
- There is no magical breakdown—just restructuring under consequence.

UED doesn't ask you to worship the unknown. It asks you to listen more carefully to what the field is still telling you.

Even at the core of a black hole—where relativity sees silence—UED sees a pulse, just too compressed to reach us... for now.

Restoring Cause and Consequence

Perhaps the most dangerous thing about the singularity is that it suspends the very thing science is built on: causality. When we say "physics breaks," we are saying: "We no longer expect continuity." We abandon the search for cause. We stop expecting a result to have a reason.

UED corrects this.

Even in the most extreme scenarios:

- Consequence continues.
- Ripple behavior evolves.
- The structured vacuum field (SVE) does not vanish—it reshapes itself under tension.

This is not mysticism. It's physics made whole again.

The Final Thought: Don't Be Fooled by the Edge

The singularity is not a holy place. It is not the origin of everything. It is not a door to magic.

It is a shadow cast by flawed extrapolation, made dramatic by our desire to assign importance to what we don't yet understand.

"The singularity was never a thing. It was the shape of our not-knowing."

And once we know better—once we observe more carefully, ripple more clearly, and model more accurately—it vanishes. Replaced not by nothing, but by structure, consequence, and coherence.

In the next section, we move from theory to transformation—from collapse to reconfiguration. We'll see how UED turns the very idea of black hole formation into a field behavior, not a fatality.

Reframing Collapse as Reconfiguration

"Nothing collapses to die. It collapses to change form."

The word collapse evokes finality—a terminal state from which nothing returns. When we hear that a star has collapsed into a black hole, we imagine destruction: form obliterated, matter lost, information erased.

This language is not only emotionally charged—it's scientifically misleading.

Because collapse, in nature, is almost never destruction. It is reconfiguration.

A leaf falls. A lung contracts. A star "dies." But in all cases, something else emerges— a shift in form, not a surrender of function.

Universal Energy Dynamics (UED) reclaims the word *collapse* from the realm of drama and death and returns it to its proper role: an energetic restructuring process. Nowhere is this reframing more vital than in our understanding of black holes.

Collapse ≠ Destruction — It Means Ripple Reorganization

In classical physics, the collapse of a massive star results in a singularity—a zero-volume point of infinite density. In this view:

- Matter is crushed beyond recognition.
- All structure vanishes.
- The region becomes a cosmic tomb.

But UED asks: *What is actually collapsing?*

Not mass. Not spacetime. But the ripple pattern that held the star's structure.

UED sees collapse as a phase transition in ripple organization:

- The stable ripple symmetries that define matter begin to fail.
- Field alignment becomes unstable.
- The system doesn't disappear—it folds into a new ripple state, one marked by tension, saturation, and misalignment.

It is not void. It is overload.

Black Holes as Storage Zones of Misalignment

Instead of singularities, UED defines black holes as regions of accumulated ripple stress:

- The field is still there.
- Energy is still conserved.
- But standard expression—light, time, structure—is suppressed by the internal conditions.

These zones serve as:

- Storage banks of unresolved ripple interaction
- Hubs of misalignment where conventional symmetry has failed
- Field reservoirs, not black boxes

Their apparent "darkness" is not because they consume all—it's because they release nothing in recognizable form until the field reorganizes.

Dynamic Fields, Not Dead Ends

A common misconception is that black holes are static endpoints. In truth, they are dynamically evolving field structures under immense pressure.

In UED, black holes are:

- Living topologies of ripple saturation
- Constantly shifting in their internal configuration
- Influenced by surrounding field changes and even by observer alignment dynamics (since perception in UED is a two-way relational process)

These fields:

- Can absorb new ripple forms
- May partially release tension over cosmic time
- React and respond to greater field equilibria in the universe

Nothing is "frozen." Everything is still in motion—just misaligned beyond our sensing thresholds.

Sites of Future Reactivation

Here is where UED makes its boldest departure:

Black holes are not permanent graves. They are potential sites of reactivation.

This does not mean matter comes back as it was. It means:

- The saturated ripple zone can, under certain field conditions, relax
- Energy can redistribute, not by explosion, but by gradual realignment
- The black hole can transition into other field forms—quietly, slowly, under new ripple logic

This is ripple evolution.

It's not hypothetical. It's the consequence of structured vacuum behavior under fatigue and restoration.

Just as pressure eventually finds a release valve, so too do black holes—through ripple exhaustion, field remapping, or interaction with larger cosmic currents.

Final Insight: Collapse Is the Field's Language of Change

What we once called "collapse" was just the field changing tone. It stopped speaking in the language of matter and light, and started speaking in the language of compressed alignment and stored consequence.

"A black hole is not where the universe ends—it's where it gathers itself before continuing."

Let's now move to the final section of this chapter: Closing Reflection: No Surprise, Just Structure — where we challenge the legacy of the singularity and embrace the truth that structure never leaves, even when our models do.

Closing Reflection: No Surprise, Just Structure

"The singularity was not the secret of the universe—it was the failure of our equations to speak its language."

For nearly a century, physicists have approached the concept of the singularity with equal parts reverence and resignation. It has been treated as a paradox wrapped in mathematics—a place where mystery reigns and where the universe itself, we are told, becomes unknowable.

But this belief was never born from observation. It was born from equation failure. From the moment general relativity's curved geometry produced infinities, we stopped looking for structure and began protecting the myth of collapse.

And yet, if there's one thing the universe has shown us consistently, it's this:

There is always structure. Even when we don't understand it. Even when our tools falter. Even when the light doesn't return.

The Universe Doesn't Hide—We Simply Don't Yet Know How to See

The idea that nature withholds its truth behind the veil of singularities is a projection of human limitation, not a reflection of cosmic design.

- The black hole core is not hidden. It's waiting for us to evolve our understanding.
- The equations didn't uncover a secret—they exposed their own limits.
- And in those limits, we drew drama instead of drawing deeper insight.

UED doesn't ask us to accept collapse. It invites us to look closer—through the lens of ripple continuity, structured vacuum behavior, and perceptual thresholds like ATT.

There was never a dead end. There was only a ripple configuration too dense to decipher with our current models.

Structure Always Exists—Even If Misunderstood

In UED, reality is fundamentally a field of structured ripple interactions. There are no voids, no breaks, no discontinuities. Only states of alignment, saturation, and transformation.

The center of a black hole is not a place where the universe disappears. It is a zone where:

- Standard matter-based coherence dissolves
- Ripple interactions enter tension saturation
- Our classical tools lose relevance

But the field continues.

There is no blank space—only active, invisible structure. Structure that still obeys cause. Still holds consequence. Still vibrates with the potential to reconfigure and re-express.

From Failure to Framework

Where relativity saw the collapse of spacetime, UED sees the reorganization of the field. Where others saw a forbidden zone, UED sees a saturated ripple chamber. Where the mainstream paused, UED continues—without assuming physics breaks, without invoking infinities, and without handing the problem off to metaphysics.

This is the defining pivot:

The singularity was not a frontier. It was a misunderstanding. A mathematical detour that turned a calculational breakdown into a cosmic icon.

We don't need to wait for a future theory of quantum gravity to patch it.

We need to reframe our relationship with perception, consequence, and the behavior of the universal field—and UED does exactly that.

Final Thought: Let Structure Have the Last Word

There are no ends in UED. Only thresholds.

What we once called a singularity was simply a location in the ripple matrix where our old definitions no longer held.

It was not a secret of the universe. It was the universe asking us to stop interpreting silence as absence.

Chapter 3: The Myth of Space Compression

The idea that space "compresses" under gravity is a geometric metaphor—useful but misleading. UED replaces it with a more fundamental explanation: field imbalance, ripple tension, and structured vacuum deformation, not spacetime curvature.

We've all seen the image: a stretched rubber sheet, sagging under the weight of a bowling ball. This is how modern physics often teaches gravity. The bowling ball represents mass. The dip in the sheet represents curved space. Smaller marbles roll toward the ball—not because they're pulled, but because the sheet itself is distorted, channeling their paths.

It's simple. Elegant. Visual. And completely misleading.

This metaphor has become a cultural icon. Textbooks love it. Animations glorify it. Even scientific papers occasionally nod to it as a gateway to deeper math.

But here's the problem:

The universe has no sheet.

There is no fabric being pulled downward. There is no external dimension stretching space like elastic. Space isn't a thing that bends—it's a coordinate system, a description of relationships between things.

And the moment we begin to take that metaphor literally, we start designing theories that treat geometry as substance—and curvature as cause.

So what if space isn't compressing at all? What if we've been watching something else entirely?

What if the phenomenon we call gravitational compression is not a folding of space, but a ripple field under stress? Not a geometric dip, but a tension zone within Structured Vacuum Energy?

What if that classic visual—the ball on the trampoline—is not what gravity *is*, but what happens when we mistake ripple imbalance for spatial deformation?

This chapter tears the rubber sheet in half. We go beneath the visual metaphor to the actual mechanics of the universe—where space doesn't stretch, and mass doesn't bend it. Instead, we find ripple density, tension accumulation, and a universe held not by curves, but by consequence.

Let's begin.

The Legacy of Compression — Spacetime in GR

"Curvature was never a force. It was a narrative — one that spoke geometry when the universe whispered ripples."

Modern gravitational theory, as framed by Einstein's General Relativity (GR), presents a radical departure from Newton's view of gravity as a force. In GR, there is no invisible pull. Instead, mass curves the geometry of spacetime, and other masses move along these curves as though they are "falling" freely through distorted terrain.

This was a conceptual triumph. It reimagined the cosmos as a four-dimensional dynamic stage—not passive emptiness, but a geometric entity responsive to mass and energy.

The fundamental slogan became:

"Mass tells space how to curve, and space tells mass how to move."

It sounded profound, and it worked. The mathematics could explain:

- The bending of starlight around the sun.
- The precession of Mercury's orbit.
- The dilation of time in high-gravity environments.
- The propagation of gravitational waves.

But beneath this success, a dangerous metaphor was quietly embedded into physics—one that described reality as bendable space instead of understanding it as interactive field behavior.

Compression in GR: Geometry as Illusion of Force

Although space cannot literally shrink or stretch (since it is not made of substance), GR treats the coordinate distances between points as changeable. When mass is present:

- The distance between two locations appears shorter than it would in flat space.
- Time slows down near massive bodies (gravitational time dilation).
- Paths of motion curve, creating the impression of attraction.

This led to a visual and conceptual model in which space seems to "compress" under gravity:

- Near Earth, clocks tick more slowly than in orbit.
- Near a black hole, time approaches zero.
- Distances between events become warped in ways that defy intuition.

To an outside observer, these effects look like compression of space and time. But in GR, this is not literal compression—it's a redefinition of measurement using the mathematical language of metric tensors and curved geodesics.

Still, the perception persists:

Heavy things shrink the space around them. Extreme masses collapse space into singularities.

This is the legacy of geometric thinking—an inheritance that UED challenges at its foundation.

Where the Metaphor Breaks: The Singularity Trap

As we move deeper into the gravitational well—toward neutron stars, then black holes, and finally singularities—the idea of compression becomes increasingly dramatic.

- Space is said to fold in on itself.
- Distances shrink to zero.
- Time is stretched so far it effectively stops.
- At the core of a black hole, geometry is said to collapse entirely, producing a point of infinite curvature and zero volume.

But what does this *really* mean?

It means that the equations governing spacetime fail to produce a coherent picture when taken to extremes. The mathematical structure of GR can no longer describe what's happening—so it describes what it can't understand as infinite distortion.

Compression becomes collapse. Collapse becomes singularity. Singularity becomes myth.

And the universe is no longer seen as structured under pressure, but as torn by gravity—a view UED finds not only incorrect, but unnecessary.

UED's Upcoming Response

UED does not require space to bend. It does not ask for time to slow. It replaces all of this with a ripple-based framework where:

- Gravity is ripple field tension, not geometric distortion.
- Collapse is ripple saturation, not curvature singularity.
- "Compression" is a visible consequence of misalignment, not a literal squeezing of dimensional space.

The next sections will explore how UED dismantles the illusion of compression and rebuilds the narrative around what's truly happening in the field: imbalance, consequence, and tension within Structured Vacuum Energy.

"Where geometry claimed collapse, ripples simply held their breath."

UED Rebuttal: Geometry Is Description, Not Mechanism

"Geometry may describe the dance—but it is not the dancer."

At the heart of General Relativity lies an elegant idea: geometry governs behavior. Mass curves spacetime, spacetime curvature determines motion. This framework replaces force with form, rendering gravity a property of bent paths rather than pulling particles.

But Universal Energy Dynamics (UED) calls for a fundamental correction to this narrative—not because geometry is wrong, but because it's mistaken for reality.

Geometry Does Not Act

UED's first principle in this critique is simple:

Geometry does not act. It represents what has already happened.

In physics, we often confuse the map for the mechanism. Coordinates are drawn after the phenomenon occurs. Curvature is calculated based on motion already observed. But this is descriptive, not causal.

- Geometry is a mathematical summary.
- It encodes the consequence of a deeper process—but not the process itself.
- It cannot generate force, tension, or acceleration.

UED shifts the origin of behavior away from curved coordinates and returns it to the field's internal ripple dynamics, which exist prior to and independent of any geometrical description.

Spacetime Is Not a Substance

This misunderstanding is deepened by how often spacetime is portrayed as a physical medium—something bendable, stretchable, even tearable. These visual metaphors have taken hold in both scientific thinking and public imagination.

But UED insists:

Spacetime is not a thing. It's a framework.

- It is a coordinate system: a way to describe positions and intervals relative to one another.
- It doesn't curve like rubber or contract like metal.
- It doesn't have stress limits, tension thresholds, or elastic behavior.

When physics says that "space is compressed" near a massive object, it's really saying:

- "Distances, as measured by observers using these coordinates, appear altered."
- "Time intervals appear to stretch or shrink in these measurements."

That's an interpretation, not a material event.

UED asserts that behind these measured distortions lies ripple field behavior—not geometry warping, but ripple misalignment and saturation within the Structured Vacuum Energy (SVE) field.

Ripple Tension, Not Dimensional Collapse

So what's really happening when space *seems* to compress?

In GR:

- The geometry becomes increasingly curved.
- The metric shrinks spatial intervals.
- Time slows.

In UED:

- Ripple tension increases as the field absorbs mass-energy

configurations.
- The SVE becomes saturated or strained.
- Alignment Threshold Time (ATT) expands, delaying interaction and perceptual coherence.

This doesn't mean dimensions are shrinking. It means:

- The field is entering a state of high-pressure ripple density.
- Observers within or near this field experience delayed alignment, incoherent signal feedback, and increased consequence latency.

The effect is not due to "compressed space," but ripple overload—like a crowded room where nothing can move easily, not because the walls have collapsed, but because the paths between participants are jammed with noise.

"Space doesn't compress. The ripple field simply resists interaction."

Why This Matters

When we say geometry acts, we cede agency to a passive model. We elevate metaphor to mechanism. But UED calls us to a higher standard:

- Understand the cause behind the curve.
- Identify the field response behind the measurement distortion.
- See space and time not as things to bend, but as relationships altered by ripple pressure.

This change in perspective eliminates the need for:

- Singularities

- Infinite curvature
- Space compression
- Time collapse

Instead, we gain a coherent, continuous understanding of how the universe responds to energy density: through ripple reconfiguration, not geometric violence.

Final Thought: The Map Cannot Shape the Terrain

Geometry, like language, is useful. But it does not create what it describes.

The compression of space is not real. The tension in the ripple field is. The collapse of dimensions is not happening. The saturation of consequence is.

"Don't ask what space is doing. Ask what the field beneath it is refusing to let go of."

Field Imbalance, Not Space Warping

"What looks like warped space is actually a stressed field crying out for realignment."

In General Relativity, mass causes spacetime to bend. But in Universal Energy Dynamics (UED), there is no such thing as "bending space." What appears as curvature in geometry is, in reality, a projection—a shadow cast by ripple dynamics within a deeper field.

UED introduces a more fundamental concept: field imbalance. This is not a metaphor. It is a physical condition in the Structured Vacuum Energy (SVE)—the all-pervading ripple-active medium that underlies everything we perceive.

What Is Field Imbalance?

Field imbalance occurs when:

- The natural, stable alignment of ripples within SVE is disturbed by excess mass-energy.
- Local ripple flows become chaotic, overcompressed, or tangled.
- The field struggles to distribute energy efficiently through alignment pathways.

In essence:

Ripple alignment fails to stabilize, and instead of self-correcting, the field becomes tense, saturated, and congested.

This condition is not abstract—it's measurable in principle through:

- Increased Alignment Threshold Time (ATT)
- Ripple density spikes
- Attenuation of signal feedback and perception

Where standard physics sees a curvature of coordinates, UED sees an active ripple overload—a pressure cooker of misalignment, not a fold in a geometrical sheet.

Ripple Congestion, Not Curvature

In GR:

- Compression appears near massive objects as shrinking distances and slowing time.
- The math suggests curved geometry: a visible effect, not an underlying cause.

In UED:

- The same phenomena are interpreted as ripple congestion.
- Nothing is being curved; instead:
    - The field is becoming saturated with unresolved ripple interactions.
    - Coherence within the ripple structure breaks down.
    - Realignment becomes slower, more difficult, or impossible.

This is why near a black hole:

- Radiation becomes attenuated.
- Information becomes unreadable.
- Time *appears* to slow.

But UED clarifies:

It's not time that's stretching. It's consequence delivery that's lagging.

The Field Isn't Shrinking — It's Saturating

A key insight of UED is this:

The field doesn't collapse. It reaches capacity.

This saturation threshold is where ripple pressure exceeds the field's ability to reorganize dynamically. Instead of snapping or bending, the field:

- Absorbs ripple tension.
- Enters a high-density ripple state.
- Suppresses interaction, not by force, but by unresolvable misalignment.

So, the "compression" near black holes isn't space being squeezed—it's the ripple matrix reaching overload, becoming too tangled to support conventional exchange of energy, motion, or even light.

This isn't gravitational collapse. This is consequence bottleneck.

Geometry Warps as a Projection of Field Behavior

Why does General Relativity appear to work so well, then?

Because geometry is a mathematical lens—an overlay we use to describe how things appear to move, not what causes them to move.

In curved geometry:

- Objects follow "geodesics," or shortest paths on a curved surface.
- These paths describe what we observe in gravitational fields—but not why it happens.

UED reveals that:

- Geodesics are the visible results of ripple-guided interactions under stress.
- The warping of geometry is a projection of ripple misalignment pressure, not the physical deformation of space.

This is like mistaking the tension lines in a stretched net for the source of the stretching. They show you what's happening, but not what created the strain.

Final Insight: Stop Watching the Shadow

We've built a century of physics on the idea that space curves under mass. But that idea was born from observing what moves, not understanding what drives the motion.

UED restores that understanding by identifying the true cause:

- Not curvature.
- Not compression.
- But field imbalance, ripple congestion, and the thresholds of alignment itself.

"You thought space was curving. But it was just the ripple field struggling to breathe."

Let's now proceed to Section: The Illusion of Shrinking Distances, where we dismantle the perception that distances contract and clocks slow because of curved space—instead, showing how delayed ripple interaction creates that illusion through field stress and perceptual thresholds.

The Illusion of Shrinking Distances

"What looks like a shortcut in space is really a slowdown in understanding."

From the lens of General Relativity (GR), the universe seems to deform itself in response to mass. Close to heavy objects—stars, planets, black holes—space is said to contract, and time dilates. The fabric of spacetime becomes warped, leading to effects like:

- Shortened radial distances toward the mass center (as in the Schwarzschild metric),
- Slower ticking clocks at lower gravitational potentials,
- And light bending around massive bodies.

This is often interpreted as physical space being squeezed, and time being stretched.

But this, according to Universal Energy Dynamics (UED), is a perception illusion, not a material transformation.

How GR Sees It

GR says:

- Gravity is not a force; it's geometry.
- Mass distorts spacetime.
- Objects simply follow geodesics—"straight" paths in this curved geometry.

From this, the following observations are drawn:

- The radial distance to a black hole becomes finite—even though classically it might appear infinite.
- Light takes longer to escape due to curvature and time dilation.
- Time "slows down" in deeper gravitational wells.

These descriptions are mathematically consistent—but they describe symptoms, not causes.

How UED Reframes It

UED replaces the curved-spacetime paradigm with something deeper:

Ripple dynamics in Structured Vacuum Energy (SVE).

In this view:

- There is no curved space.

- There is only ripple behavior—the motion and tension of consequence across a structured, ripple-capable vacuum.

So what appears as:

Shrinking distance near a massive object is actually:

- A failure of ripple alignment—the path is still there, but the ripple can no longer traverse it efficiently.
- The route becomes functionally inaccessible, not physically shorter.

Slower time is not time stretching—it's:

- A delay in ripple feedback due to high local tension.
- ATT (Alignment Threshold Time) expanding, meaning that response time to stimuli—light, motion, perception—lengthens, not time itself.

Thus:

The distance hasn't changed. The field's ability to align across that distance has.

ATT and Perception Lag

Let's revisit ATT—Alignment Threshold Time, the time it takes for an observer to successfully perceive a ripple realignment from the field.

In low-tension environments:

- ATT is minimal.
- Light, motion, and causality appear normal.

In high-tension zones near massive bodies:

- Ripple stress increases.
- ATT increases—sometimes to functional infinity.
- As a result:
    - Clocks appear to "slow" (they're not slower—feedback is delayed).
    - Distances appear to shrink (they're not shorter—alignment pathways are just congested).

This is not gravitational time dilation—it's field strain manifesting as delayed interaction.

No Shortcuts, Just Misalignment

In GR, paths curve and compress around mass. But in UED:

- No path is shorter.
- No time is slower.

What changes is your ability to resonate with the field.

"A road isn't gone because it's foggy. It's just not visible—yet still there."

The illusion of shrinkage is caused by perception failure, not physical change.

That's why light appears to bend—it's not navigating geometry, it's finding ripple paths distorted by tension. That's why motion near a black hole seems sluggish—not because it's slow, but because your frame of alignment can no longer process the ripples in real-time.

Closing Insight: Shrinkage Is Not a Physical Event

The GR picture gave us beautiful equations to describe warped space. But UED invites us to discard the metaphor and ask a more grounded question:

"What is really failing here—space, or our relationship to it?"

The answer: Our ripple interaction fails under tension. And what we interpret as shrinking, slowing, or bending—is just the visible signature of misalignment.

"Space doesn't compress. Our ability to meet it does."

Visual Metaphors vs. Physical Realities

"The trampoline was a teaching tool—not a blueprint of the cosmos."

For decades, the most popular way to visualize gravity has been the trampoline analogy—a large elastic sheet representing space, sagging under the weight of a heavy ball that symbolizes mass. Small balls placed on the sheet roll toward the depression, mimicking how smaller objects move toward larger ones in space.

It's clever. It's memorable. It's also profoundly misleading.

The Flaws in the Trampoline Analogy

Let's dissect this metaphor carefully.

1. It assumes an external force pulling downward The trampoline sags because gravity on Earth pulls the mass down. But in the analogy, gravity is the very thing it's supposed to explain. This creates a circular problem:

You can't use gravity to explain gravity.

1. It treats space as a rubber-like substance The metaphor

assumes space can stretch, compress, and bend like an elastic membrane. But what exactly is stretching in the vacuum of space? What are the material properties of this "fabric"?

2. It implies curvature is caused by weight, not by interaction The sheet doesn't show the internal dynamics of the universe—it shows a 2D simplification of motion paths. It hides the real question:

Why does mass influence space in the first place?

UED: From Fabric to Field

In Universal Energy Dynamics (UED), this analogy is not only obsolete—it's replaced by a ripple-based model grounded in observable consequences.

Here's how UED reframes the idea:

- Space is not a fabric There is no elastic substance. What exists is Structured Vacuum Energy (SVE)—a ripple-active field that holds the potential for all interaction.
- Mass is not an external object "placed" on the field Mass arises as ripple configurations within the SVE itself. It is not something acting on the field; it is of the field.
- There is no "pull" from curvature What appears as gravitational attraction is, in reality, a tension response in the ripple matrix.

Thus:

"Nothing pulls. Everything adjusts to restore balance."

Gravity, in this view, is not a force or even a passive geodesic—it is a ripple realignment seeking equilibrium in a disturbed field.

Let's Drop the Sheet

The trampoline metaphor is comforting because it's visual and tactile. But it makes dangerous assumptions:

- That space has directionality (e.g., "downward" curvature)
- That mass acts externally on the field
- That space can behave like a surface

UED calls this out clearly:

- There is no surface
- There is no external direction
- There is only internal tension and dynamic realignment

What we observe as curvature is a projection of how ripples redistribute to accommodate high-density configurations.

No dips. No sheets. Just fluctuations in the rhythm of the field.

Final Insight: Metaphors Teach, But They Must Evolve

The trampoline metaphor helped a generation imagine invisible forces. But it also locked us into a geometry-first worldview, blinding us to deeper mechanisms.

"We gave the universe a sheet to sleep on— But forgot to ask what makes it move."

UED replaces metaphor with mechanism. It gives us ripple logic, field tension, and observer-attuned perception—tools that not only describe reality but emerge from it.

Why Compression Narratives Create Conceptual Dead Ends

"You can't shrink what never had size. You can't fold what was never flat."

The more we try to squeeze meaning out of "space compression," the more we drift into paradoxes. What begins as a useful analogy ends in conceptual disaster when taken literally. Let's explore why.

The Fatal Assumption: That Space Is Something

At the heart of the compression narrative is one unspoken assumption:

Space exists as a real, stretchable substance.

But this has never been proven. In fact:

- Space has no observable texture, no density, no elastic properties.
- It has no intrinsic direction, no mass, no form.
- Its only known function is to provide reference between things—not to *be* something itself.

Yet, the language of modern physics often treats space as if it were made of rubber:

- "Space stretches with the expansion of the universe."
- "Space compresses near black holes."
- "Space is curved by mass."

This poetic language becomes problematic when embedded into mathematics that try to model extreme conditions—especially near singularities.

The Result: Mathematical Collapse

When you assume space can be compressed, your equations allow it to do so—sometimes infinitely. This leads to:

Singularities, Points where:

- Density becomes infinite
- Curvature becomes infinite
- Physics ceases to offer any prediction

Undefined Boundary Conditions

- What happens *at* the singularity?
- Is there a center? A core? A dimensionless point?

Conceptual Contradictions

- If space has no substance, how does it compress?
- If time slows infinitely, is it still time?
- If distances shrink to zero, does direction matter?

These aren't mysteries to be solved—they're logical traps created by flawed premises.

UED's Liberation: Don't Compress—Realign

Universal Energy Dynamics (UED) sidesteps these dead ends by refusing to treat space as a compressible entity. In UED:

- There is no "space" to compress. What we call "space" is just the structured vacuum field—a ripple-ready medium, not an object.
- What changes is not space—but field tension. The field saturates, misaligns, and distorts its ripple structure. But it doesn't "compress" in a geometric sense.

- Collapse is reinterpreted as ripple overload. Singularities become zones of maximal misalignment, not infinite density.

This removes the need for:

- "Infinite curvature"
- "Zero-volume points"
- "Broken laws of physics"

From Dead Ends to Dynamic Tension

By moving away from compression-based thinking, UED reframes black holes, extreme gravity, and even cosmology as expressions of ripple behavior—not as geometric deformities.

This allows:

- Continuity of physics even in extreme zones
- A path for interaction and observation based on ATT (Alignment Threshold Time)
- A consistent mechanism that doesn't break under pressure—literally or mathematically

"Nature doesn't collapse. Our descriptions do."

◈ Closing Insight: Let Go of Compression, Find Clarity

The compression narrative served us when we needed visual metaphors. But to move forward—to build a physics that aligns with nature instead of metaphor—we must let it go.

"What we thought was a crushed space... Was just a field trying to breathe."

The UED Model: From Curvature to Ripple Topology

"The universe isn't bent—it's tense. And tension doesn't warp—it realigns."

The classical model of General Relativity gave us a poetic image: A universe shaped by curves. Mass bends spacetime, and everything else follows that curvature.

But as we've explored in previous sections, this is a model, not a mechanism. A description, not a cause.

It's here that Universal Energy Dynamics (UED) offers a decisive shift—from a geometry-dominant worldview to a ripple-topological model rooted in field behavior, tension distribution, and consequence-based alignment.

Let's break down this shift.

1. From Curved Spacetime to Ripple Flow Fields

- **GR View:** Space and time form a smooth fabric. Mass distorts this fabric. Objects follow the curves.
- **UED View:** The universe is a ripple-responsive field—Structured Vacuum Energy (SVE). What we interpret as "curvature" is actually ripple flow topology:
    - The direction and intensity of ripple propagation
    - The field's internal stress-response to embedded ripple mass-forms

Ripple flow doesn't "curve" a coordinate system—it reshapes interaction zones.

2. From Spatial Compression to Field Saturation

- **GR suggests:** Near massive bodies, space compresses and time slows.
- **UED reveals:** Mass induces localized ripple congestion. The field saturates, not shrinks.

➤ Think of it as traffic congestion in a ripple field:

- Ripples struggle to realign efficiently.
- This creates perceptual delays (higher ATT).
- The environment feels "compressed" because feedback is throttled.

"Space doesn't compress. It clogs with unresolved tension."

3. From Geodesics to Alignment-Based Response

- **In GR:** Objects follow geodesics—the shortest path in curved spacetime.
- **In UED:** Objects respond to ripple gradients—they move in the direction of least resistance in the ripple matrix.

This shift is profound:

- No need to invoke curvature equations.
- No dependence on predefined spacetime geometry.
- Motion is a consequence of field dynamics, not a traveler obeying invisible grooves.

Unified Insight: Observables Reinterpreted

Everything we previously modeled as effects of curvature or compression can now be traced back to ripple interactions under stress:

Classical Effect (GR)	UED Reframe
Space curvature	Ripple field tension
Time dilation	ATT extension (delay in perception)
Spatial compression	Saturation of ripple density
Free-fall trajectory	Realignment through ripple gradients
Event horizon	ATT threshold for field resonance

This not only offers mechanical clarity, but restores physical continuity where traditional models break down.

Closing Insight: From Geometry to Consequence

What Einstein modeled as a curved map, UED reveals as an active terrain—flexing not in dimensions, but in ripple pressure.

"The universe is not shaped by curves—but by responses." "It doesn't bend. It breathes."

We don't need compressed space. We need field-aware models. And UED gives us exactly that.

Implications for Astrophysics and Cosmology

"Remove compression, and the universe becomes clearer—not smaller."

By replacing the idea of curved, stretchable spacetime with ripple-based field dynamics, Universal Energy Dynamics (UED) rewrites some of the most foundational assumptions in modern cosmology and astrophysics.

This is not a minor theoretical adjustment. It's a structural realignment of how we view reality—from fabric metaphors to field consequences. Let's now explore what this means for major concepts in the universe.

1. Black Hole Cores: Not Compressed Voids, but Saturated Fields

GR Interpretation:

- At the heart of a black hole lies a singularity—an infinitely small, infinitely dense point.
- All mass "compresses" into this void, breaking the laws of physics.

UED Reinterpretation:

- The black hole core is not a compressed object—it is a zone of ripple saturation within the Structured Vacuum Energy (SVE).
- No point of infinite density.
- No "void."
- Instead, it's a dense misalignment region, where ripple feedback cannot occur in normal time (i.e., ATT tends toward infinity).

"Collapse was a misunderstanding. Saturation is what really happens."

2. The Early Universe: Not a Bang from Compression

Big Bang Legacy:

- The universe began as a singularity: a point of infinite density and zero volume.
- Space and time "expanded" from this initial compressed state.

UED Response:

- The universe did not emerge from compression—it manifested from a shift in field balance.
- The so-called "beginning" was a phase transition in SVE:
    - From stable equilibrium to dynamic ripple propagation.
    - Energy didn't explode—it rippled into structured existence.

This redefines the origin not as a *blast* from nowhere, but as a coherent emergence from a pre-existing ripple-neutral field.

3. Gravitational Waves: Not Space Stretching—Ripple Distortions

Mainstream View:

- Gravitational waves are ripples in spacetime itself, stretching and squeezing space as they move.

UED Perspective:

- These waves are modulations of ripple tension in the SVE field.
- What detectors like LIGO capture is not "space stretching" but field distortions that temporarily alter alignment thresholds.

The signal isn't a wave through space—it's a fluctuation in ripple density and responsiveness, observable as displacement or timing shifts in instruments.

4. Distance in the Universe: A Function of Ripple Latency

GR/Standard Cosmology:

- As the universe expands, space itself stretches.
- Galaxies are "moving apart" as the fabric of space grows between them.

UED Clarification:

- Space does not stretch. It doesn't need to.
- What increases is the latency of ripple response—the time it takes for alignment between distant regions.

This redefines cosmic distance as relational ripple delay, not expansion of coordinates.

For instance:

- Redshift isn't proof of space stretching.
- It's an indicator of decreasing ripple synchrony over vast separations.

"The universe isn't growing—it's desynchronizing."

Summary: Reconstructing Reality with Ripple Logic

Astrophysical Concept	GR Interpretation	UED Interpretation
Black Hole Core	Infinite compression	Ripple saturation zone
Big Bang	Singularity expanding into space	Field transition triggering ripple emergence
Gravitational Waves	Space-stretching ripples	Tension distortions in SVE
Cosmic Distance	Coordinate expansion	Ripple latency and field delay

This shift removes the need for speculative metaphysics (singularities, voids, dimensional tears) and grounds cosmic behavior in observable, testable field responses.

Final Thought: From Expansion to Expression

By discarding compression as a real phenomenon, UED does more than simplify equations—it liberates our imagination.

"The cosmos didn't erupt from nothing. It responded to a change in tension."

"We don't measure distances. We measure dissonance."

This isn't reduction—it's refinement. And it lays the groundwork for a more coherent, consequence-based understanding of the universe.

Closing Reflection: You Were Never in a Curved Room

"The universe never folded in on itself. It simply lost its rhythm long enough for us to call it curvature."

You Were Never in a Curved Room.

Not in your childhood stargazing. Not in the telescopic images of galaxies twisted by invisible hands. Not in the equations that bent space to fit our ignorance. And not in the physics class where the trampoline became our temple.

You were never in a room made of curves. You were always in a ripple field—a universe alive with tension and release, harmony and resistance. What you mistook for curvature... was consequence.

The Rubber Sheet Was a Metaphor—Nothing More

It helped us imagine. But then it began to rule us. We forgot that space isn't a fabric and gravity isn't a weight. We forgot that descriptions are not things—and metaphors aren't mechanics.

The truth?

- Space never compressed.
- It never folded.
- It never snapped back like elastic.

Instead, it responded—to every particle, to every ripple, to every imbalance that sought alignment in the Structured Vacuum Energy (SVE). It didn't bend. It breathed.

The Illusion of Curvature Was a Symptom of Misalignment

Why did we see curved paths? Why did clocks tick differently? Why did light bend?

Because ripple tension altered the conditions for perception. Because ATT changed. Because the field was out of tune—not spatially, but energetically.

We thought distances had shortened. But it was our resonance that stretched thin. We thought time had slowed. But it was our feedback loop that weakened.

We mistook the song of a strained violin for the shape of the room it played in.

UED's Parting Wisdom for This Chapter

- The universe doesn't curve—it coordinates.
- It doesn't compress—it saturates.
- It doesn't expand—it rebalances.
- It doesn't fold—it re-tunes.

The more we honor consequence over construction, the more this truth becomes obvious:

We never needed space to bend. We only needed to listen to the field.

Closing Quote

"The universe never folded in on itself. It simply lost its rhythm long enough for us to call it curvature."

Interlude

Beyond the Event, Beneath the Illusion

"If the horizon was never a wall, and the singularity never a point, then what exactly have we been chasing?"

We have peeled back the first layer of illusion.

The Event Horizon—once seen as the edge of no return—is no longer a cosmic trapdoor. It is a threshold of perceptual failure. It is not a structure, but a stall in ripple alignment—a place where our feedback loop with the universe fails to complete itself.

The Singularity—revered as a place where physics ends—is no longer an abyss of ignorance. It is a mathematical panic button, triggered when our models hit their own limits. UED replaces this with the idea of ripple saturation—a condition, not a catastrophe.

And the Curved Space we once bent our understanding around? That was never a feature of the universe. It was the shape of our misunderstanding, projected onto the field.

What emerges from these chapters is not just a new physics—but a new philosophical posture.

One that does not seek the fabric of space, but the consequences of imbalance. One that does not treat gravity as a pull, but as ripples responding to tension. One that stops asking, *"What is space made of?"* and starts asking,

"What is the universe trying to restore?"

So, where do we go from here?

If Part I dismantled the prison of the black hole—Then the next step is to explore what moves, what bends, and what echoes across the cosmos without needing curvature, compression, or collapse.

We now turn our gaze to the field itself—To what ripples carry, what waves distort, and what light truly means when it no longer travels, but simply aligns.

Prepare to leave the black hole behind.

What comes next is the reconstruction of everything we thought was torn.

Part II: Information, Radiation & Repercussions

Chapter 4: Information Never Dies — It Resonates

What appears lost at the edge of a black hole isn't destruction—it's realignment. Information is not erased; it's displaced, encoded, and conserved in ripple form within Structured Vacuum Energy (SVE).

Imagine taking a beloved book—one filled with stories, memories, and meaning—and setting it on fire. The pages curl, the ink smears into flame, and eventually, there's nothing left but ash, heat, and smoke. It's easy to say the book is gone. But is it really?

That fire didn't annihilate the story—it redistributed it. The light carried the pattern of the print. The heat held the energy of its bonds breaking. The smoke released its molecular memory into the air. The book's form disappeared, yes. But its information simply changed its mode of presence.

If we accept that energy is conserved, then why wouldn't information be?

This was the question that haunted Stephen Hawking for decades—though he asked it in the language of black holes. His dilemma was not poetic. It was brutal.

"Black holes," he warned, "are not as black as they are painted."

They radiate. They evaporate. They die. And yet—nothing that fell into them ever came back out. Not as form. Not as signal. Not even as trace.

Thus emerged one of modern physics' most disturbing puzzles:

The Information Paradox.

A contradiction so severe that it tore at the foundation of quantum theory and made a villain out of gravity itself.

If information could truly vanish behind an event horizon, then the universe itself forgets. It loses continuity. It breaks causality. It permits a kind of cosmic amnesia.

But what if that's not what's happening?

What if the black hole never destroyed the book— It simply translated it into a language we hadn't yet learned to hear?

What if Hawking's paradox was never about loss, but about misunderstood resonance?

This chapter begins with a simple but powerful proposal: Information never dies. It ripples. It echoes. It realigns. And in the structure of the vacuum itself, every memory waits to be heard again.

What Hawking Proposed: The Information Paradox

In the 1970s, Stephen Hawking revolutionized our understanding of black holes by daring to suggest they weren't eternal. Contrary to their fearsome reputation as bottomless pits from which nothing escapes—not even light—Hawking showed that black holes radiate. This now-famous prediction, known as Hawking radiation, emerged from combining principles of quantum mechanics with the gravitational curvature of spacetime.

According to his theory, quantum fluctuations near the event horizon constantly give rise to virtual particle-antiparticle pairs. Occasionally, one particle escapes while the other falls into the black hole. To an outside observer, the black hole seems to emit energy—a slow leak of radiation that continues over cosmic timescales.

But here's the catch: this radiation is thermal. It carries no information about what fell into the black hole. Whether it was a rock, a human, or an encyclopedia, the outgoing radiation looks the same—just heat, random and structureless.

This simple insight had a devastating implication.

In the world of quantum mechanics, one principle reigns supreme:

Unitarity — the idea that information is never truly lost. The future of a quantum system is fully determined by its past, and conversely, the past can be reconstructed from the future. The wavefunction evolves predictably—it does not vanish.

Hawking's proposal violated this sacred law. His radiation seemed to erase the past. The black hole evaporates, and the information—encoded in whatever it had swallowed—goes with it, never to be retrieved.

This is what became known as the Information Paradox. If black holes permanently delete information, then quantum theory breaks. And if quantum theory breaks, then predictability, causality, and even logical consistency in physics start to crumble.

Hawking captured the philosophical depth of this conundrum with sharp wit: "God not only plays dice, but sometimes throws them where they can't be seen."

What he meant was: randomness isn't the issue—opacity is. It's not enough that the universe behaves probabilistically. If those probabilities no longer preserve the fingerprint of past events, then the book of reality is being rewritten with missing pages.

This led to decades of debate, controversy, and increasingly elaborate theoretical gymnastics to "save" the information without discarding Hawking's radiation mechanism. Solutions ranged from holographic principles to firewalls to quantum entanglement bridges—creative, complex, and still incomplete.

But perhaps the real problem wasn't in rescuing the information from destruction. Perhaps the real problem was in how we understood where it lived in the first place. Was information ever "inside" the black hole as a location—or was it always part of a larger field interaction, never isolated, never gone, only resonating in forms we've yet to decode?

In the next section, UED offers a radical pivot—not by patching Hawking's equations, but by rewriting the substrate in which information exists.

The Flawed Assumption: Information as Form, Not Field

At the heart of Hawking's Information Paradox lies a subtle but critical error—an ontological misplacement of information itself.

Contemporary physics, especially in the wake of classical mechanics and general relativity, has been obsessed with form. Whether it's particles, waves, or fields, we've been conditioned to associate information with structured form—the arrangement of matter, the configuration of quantum states, or the pattern inscribed in physical coordinates.

This mindset works well for conventional systems. We can read data from a hard drive, measure spin states, trace particle interactions—all with the assumption that form carries meaning. When that form is destroyed—say, by fire, disintegration, or gravitational collapse—so too, we think, is the information it encoded.

This is where the paradox takes root. If an object falls into a black hole and vanishes beyond the event horizon, and if Hawking radiation reveals no imprint of what fell in, then by the logic of form-dependent information, that information is gone.

But this is a category error.

It confuses visibility with existence, and structure with state.

In Universal Energy Dynamics (UED), information is not defined by persistent form, but by field resonance—the ripple patterns generated by any object's interaction with the Structured Vacuum Energy (SVE). Every action, every configuration, every collapse creates disturbances, alignments, and resonances in the field. These don't vanish—they dissipate, transform, or remain latent depending on the field conditions.

To assume that destruction of form equals loss of information is to ignore how memory works in nature. Consider:

A stone dropped in water disappears beneath the surface—but the ripples tell you it was there.

A voice echoes across a canyon—even if the speaker is long gone.

Burn a book, and the light, heat, smoke, and ash still contain clues—if not the exact syntax, then at least the energy and interaction history of what once existed.

In each case, form dissolves, but effect persists. And that persistence is not in material continuity, but in field signature—an imprint on the medium itself.

This is where Hawking's logic falters.

His equations assume that information is a thing, tied to matter, stored locally, and annihilated upon crossing a point-of-no-return.

But UED sees information as a relational imprint, created at the moment of interaction and immediately diffused into the ripple field.

In this view, the black hole does not delete information—it redistributes its signature. The moment something engages with extreme ripple tension near the event horizon, it alters the field, permanently. The SVE remembers, not because it stores "files," but because its very structure is defined by every ripple it has ever processed.

Thus, the paradox isn't that information is lost. The paradox is that we ever believed it had to be inside something to exist.

Transition to Next Section: What if form was only the first layer of memory—and the field, not the particle, was the true archivist of reality? Let us now explore how UED reconceives information—not as encoded symbols—but as resonance states etched into the structured vacuum itself.

UED's Core Reframe: Information as Ripple Pattern

To resolve the information paradox without breaking causality, we must stop asking *"Where did the information go?"* and start asking *"How does the field respond when form changes?"*

Universal Energy Dynamics (UED) offers a complete reframe: Information is not a stored object—it is a ripple configuration.

Let's unpack this.

In classical thinking, information is treated as if it resides within particles, inside structures, or on surfaces—stored like code on a hard disk or ink on paper. This model assumes that things contain meaning.

But UED dismantles this. In ripple-based physics, nothing exists in isolation. Every entity—whether a photon, an atom, or a collapsing star—is a ripple event, a disturbance in the Structured Vacuum Energy (SVE). What we observe as "object" is simply a stable ripple pattern—coherence in the field.

Thus, information is the pattern, not the object. It's the arrangement of ripple behavior, the interference landscape, the relational structure of tension, motion, and alignment across the field.

Think of a pebble dropped into a calm pond. The stone vanishes beneath the surface—but its history spreads outward as concentric waves. These ripples are not the object—but they are the record of the event.

UED generalizes this idea:

- When something happens—a star collapses, a photon scatters, or a particle decays—it causes ripple shifts in the surrounding field.
- These shifts alter the alignment conditions of the SVE.
- The field reorganizes to accommodate the new imbalance—and that reorganization is the memory.

Even if the form is destroyed—whether by black hole collapse, quantum decay, or thermodynamic diffusion—the field retains the resonant imprint of what occurred.

This memory is not local. It's holographic.

Just as every piece of a hologram contains the whole image (though with reduced clarity), the ripple field stores information redundantly. It spreads, weakens, and blends—but it does not vanish. The SVE acts as a distributed ledger of all interactions, storing not data points, but relational distortions that evolve with the field.

Collapse doesn't erase information—it compresses it, distorts it, and embeds it across the ripple web. Like a chord played on a string, the note fades, but its vibration resonates within the instrument—and sometimes even the room.

This is why the UED view considers black holes not as terminators, but as tension zones—regions of such intense ripple congestion that conventional realignment (and thus radiation) slows to near-zero. Yet even here, the informational consequence continues to echo across the larger field.

In UED:

- Information is never static—it is dynamic interference.
- It isn't stored *inside* particles—it emerges between them, through field interaction.
- No collapse is final—every distortion etches a new ripple memory.

This idea unlocks an elegant resolution to the paradox Hawking posed: The black hole doesn't need to return information in the form it received it. It needs only to ensure that the field it disturbed carries forward the story of its past.

And it always does.

In the next section, we'll dive deeper into this holographic principle—not as a metaphor, but as a fundamental ripple behavior in UED's architecture of the cosmos.

Ripple Memory: How the Field Remembers

Let us go deeper now—not just into the theory, but into the fundamental memory architecture of reality.

Universal Energy Dynamics (UED) asserts that memory is not exclusive to brains, devices, or storage systems. It is a natural consequence of interaction. When anything acts, aligns, or collides, it disturbs the field. And that disturbance doesn't just vanish—it becomes part of the very fabric of the Structured Vacuum Energy (SVE).

This is Ripple Memory.

Ripple memory is not a metaphor. It is a functional behavior of the field—a physical imprint made not by marking a surface, but by shaping the tension landscape of the cosmos.

Holographic – Every Part Contains the Whole

In UED, ripple memory is fundamentally holographic in nature. That is:

- Each interaction causes a distributed ripple disturbance, not a point-specific change.
- These ripples propagate through the SVE and interfere with existing ripple configurations.
- As a result, every part of the field carries a partial echo of every event.

This is not speculative—it's an energetic necessity in a medium where all effects are relational.

Think of the universe as a spiderweb in the dark. One tug—no matter how small—vibrates the entire mesh. You may not see the spider. But the web *knows*.

The ripple doesn't travel as a particle. It reorganizes tension, and that reorganization has consequences across the entire field.

Non-Local – Spread, Not Stored

Unlike classical memory—which implies a container holding a fixed value—ripple memory is non-local and embedded. It's not "saved" in a single spot. It lives in the geometry of the field.

When a massive body collapses into a black hole, UED holds that the field around it becomes saturated with that history. Even if the object is gone, the ripple tensions, distortions, and delays persist.

- This is why gravitational lensing, frame dragging, and time dilation appear even without visible sources.
- The field remembers—not by holding objects, but by retaining their consequences.

This is a new paradigm of memory:

- No container.
- No retrievable "files."
- Only distributed influence, ready to be reactivated through resonance.

Responsive – Recoverable Under Realignment

Now, perhaps most strikingly: ripple memory is not frozen—it is responsive.

That is, under correct alignment conditions, memory can resonate back into perceptible form.

- Just as a tuning fork can cause another across the room to vibrate,
- Just as trauma resurfaces through similar emotional frequencies,
- Just as past music can be reconstructed from surviving harmonics,

—ripple memory responds to matching patterns.

This is not science fiction. It is the mechanism behind recall, emergence, and reorganization. The field doesn't need to store data in bits and bytes—it simply needs to resonate under the right conditions.

In human experience, this is akin to déjà vu—not the repetition of an event, but the realignment with an old field imprint.

Residual Echoes: Memory Through Ripple Trauma

Here we introduce a powerful concept in UED: Residual Echoes.

When a ripple event is too intense—too fast, too massive, or too misaligned—the field cannot fully realign. This creates a scar, a zone of persistent ripple tension.

- These residual echoes are not passive—they affect future alignment attempts.
- They slow down ATT (Alignment Threshold Time).
- They distort local dynamics.

- And crucially, they preserve a field-level record of the trauma.

Black holes are not just gravitational anomalies—they are ripple echo chambers.

They don't "store" information in the traditional sense. They trap misalignment—and that trapped field pressure is the memory of the original event.

This is why even after evaporation (if it occurs), the surrounding field remains changed. Hawking's paradox dissolves, because the memory was never inside the mass—it was in the field all along.

In the next section, we'll explore how Hawking Radiation, far from being a random leak, may itself be a slow ripple bleed-off of distorted alignment—a process not of loss, but of eventual field restitution.

Radiation Without Loss: UED's Alternative to Hawking Radiation

Stephen Hawking's groundbreaking proposition in the 1970s—that black holes radiate energy and thus can eventually evaporate—shook the very core of theoretical physics. It fused quantum field theory, general relativity, and thermodynamics into one elegant yet unsettling prediction: black holes are not eternal. They leak.

But what do they leak? And at what cost?

In the traditional picture, Hawking Radiation arises from quantum fluctuations near the event horizon. One particle falls in, the other escapes. The black hole loses mass. Over time, it fades. Elegant? Yes. But deeply disturbing. Because the escaping radiation is thermal, not specific. It carries no memory of what fell in. Thus, the Information Paradox was born.

UED doesn't reject radiation—it redefines it.

Radiation = Ripple Discharge, Not Particle Emission

In Universal Energy Dynamics, radiation is not a discrete emission of particles from a boundary. That picture is based on outdated assumptions: particles as things, spacetime as a stage, and boundaries as physical barriers.

Instead, radiation is the field's ripple discharge—a dynamic process where tension accumulated in Structured Vacuum Energy (SVE) is gradually resolved outward through ripple realignment.

A black hole, in this view, is not a container leaking mass. It is a high-pressure ripple zone that releases stored misalignment as the surrounding field becomes capable of absorbing it.

- No singularity explosion.
- No particle-pair magic.
- Just ripple fatigue resolving itself.

The black hole doesn't *emit*. It decompresses—slowly, relationally, and responsively.

Information Reforms, Not Re-forms

Here lies the most vital distinction: in UED, information is not re-formed as particles. It doesn't return like a letter lost in the mail. It re-forms as a reconfiguration of the field.

The misaligned zone—that scarred ripple tension—begins to offload pressure. As it does, it affects the surrounding SVE, distorting ripple pathways, altering local ATT, and imprinting traceable patterns.

This means:

- What seems like featureless radiation is in fact ripple-coded.
- It carries the signature of misalignment, the rhythm of trauma, the echoes of collapse.
- To the right observer—one calibrated not to particles but to ripple topology—this discharge isn't noise. It's history in resonance.

Beyond Thermal: Signature Distortions in the Field

Hawking's model interprets black hole evaporation as purely thermal—which, in quantum mechanics, suggests randomness, information loss, and entropic ambiguity.

UED counters with a powerful alternative: Ripple emissions are not random. They're patterned, but complex—so intricately embedded in field behavior that classical observers misread them as noise.

- A black hole's "glow" may appear uniform—but it subtly modulates with the history of the collapse.
- Each emission is shaped by local ripple fatigue, not universal entropy.
- Field distortion is never uniform—it retains the signature of what caused it.

This leads to a profound insight:

In UED, what you call "random" is just temporarily unreadable alignment.

Nothing Is Truly Random—Only Delayed Alignment

At the heart of UED is the idea that randomness is an illusion born of ignorance. When you can't perceive the full ripple configuration—or when alignment takes too long to emerge—you assume chaos.

But nature is not chaotic. It is consequential.

The field does not forget. It remembers in delay.

- Radiation from black holes is not "pure noise."
- It's the language of a memory trying to re-synchronize with a calmer field.
- If we could map the ripple delays, tensions, and resonant trajectories, we wouldn't see heat—we'd see a symphony of encoded experience.

"Nothing in UED evaporates into nothing. It simply realigns when the field can bear it."

So where does this take us next?

If radiation is ripple exhaustion—not energy loss—then the very nature of black hole thermodynamics must be rewritten. The entropy of a black hole is not a measure of randomness—it's a measure of stored ripple pressure. A memory budget, not a death clock.

And what of the famed holographic principle, which claims that all information about a black hole's interior is stored on its surface?

UED sees a different canvas—not a surface, but a field in tension. Not a hologram—but a ripple record, stored across SVE.

Repercussions: The Echoes of Everything

In Hawking's vision of black holes, the end is silence—evaporation, disappearance, the loss of identity behind a veil of thermal blur. Information dissolves into entropy. The final consequence: unrecoverable history.

UED challenges this narrative not with optimism, but with principled coherence. At the heart of this challenge lies a singular premise:

Nothing ever truly vanishes—because everything leaves a ripple.

In Universal Energy Dynamics, reality is not a collection of objects but a web of interconnected ripple patterns. Every action, every change, every interaction with the field generates a consequence—not always seen, but always felt by the SVE.

These consequences form the echoes of everything.

The Universal Law of Consequence (LOC)

In UED, reality unfolds under one governing truth: The Law of Consequence (LOC). This law does not rely on force, motion, or quantum states. It is the field's guarantee that:

- Every ripple has an origin.
- Every origin leaves a signature.
- Every signature, distorted or delayed, remains encoded in SVE.

Even under extreme conditions—black holes, high-energy collapses, entropy spikes—LOC ensures continuity of effect.

Unlike GR, which allows for "event cutoffs" like horizons and singularities, and unlike traditional quantum theory, which flirts with randomness and decoherence, UED guarantees perceptual causality through ripple logic.

Even if the path is broken, the echo survives.

Information Lost? No—Only Misaligned.

Let's revisit the core misunderstanding that birthed the Information Paradox: Equating destruction with deletion.

- In GR and Hawking's models, once something crosses the event horizon, its identity is irretrievable.
- In UED, once something misaligns beyond perceptual threshold, its identity is obscured, not erased.

A lost file isn't deleted—it's corrupted, displaced, or scrambled in unfamiliar syntax.

Black holes do not delete information. They misalign it.

- The ripple configuration becomes too chaotic for standard observers.
- The ATT (Alignment Threshold Time) required to resolve these patterns stretches beyond practical interaction.
- But given the right frame, rhythm, and resonance, the field can re-cohere.

In essence, what's gone is not gone—it's just waiting for a better listener.

Field Realignment = Restoration

So how do we retrieve what appears lost?

UED proposes a compelling mechanism: field realignment.

Think of ripple distortion like a blurry image on a camera. The detail is still present—but out of phase. If you adjust the focus lens, the image returns—not from nowhere, but from within the field's own latency.

The same logic applies at cosmic scales.

- If we realign to the original ripple sequence—even partially—we can decode what was buried in the tension.
- Information stored as residual field memory can be reconstructed through ripple synchronization.
- This doesn't require time reversal—only ripple phase reconstruction.

Hence, the field is not a passive recorder. It is a living memory system.

Memory in UED is not what you hold. It is what the field holds when your alignment returns.

The Ripple Echo: Intelligence of Consequence

From consciousness to cosmic collapse, the ripple principle remains unwavering:

- You act — the field responds.
- You disappear — the field remembers.
- You collapse — the field stores the signature.

This is why UED doesn't just explain physics—it reorients philosophy.

Because if every consequence ripples eternally, then:

- Justice is not moral—it's mechanical.
- Karma is not belief—it's field response.
- Awareness is not illusion—it's real-time ripple reflection.

Hawking's universe left us with evaporated objects and lost truths.

UED's universe leaves nothing behind—because everything is in front of us, waiting for resonance.

What does it mean to "encode everything" in the field?

What does this say about black hole entropy, quantum holography, and the boundary-based view of physics?

We now move from information's survival to its representation.

Next up: how ripple fields give rise to holographic realities—and why UED suggests no surface can hold what the field doesn't ripple.

Closing Quote

"Nothing in the universe is ever forgotten—only misaligned. To find truth again, you don't dig deeper. You tune finer."

The Death of the Paradox

The paradox dies not with answers, but with better questions. In UED, we don't solve the paradox of black hole information loss—we dissolve it. Not with complexity, but with clarity.

Because what if the paradox never belonged to the universe at all?

What if it was always ours—rooted in flawed assumptions about space, matter, and the meaning of information?

Assumption 1: Space as a Container

Stephen Hawking's entire framework—along with classical and relativistic physics—treats space like a cosmic box. Things fall into it, cross boundaries in it, vanish from it. And when those things vanish, the question becomes: *Where did they go?*

UED offers a clean correction:

Space isn't a container. It's a consequence.

There is no "inside" or "outside" of space—only zones of ripple activity within Structured Vacuum Energy (SVE). In this model:

- A black hole isn't a pit.
- An event horizon isn't a lid.
- "Falling in" is just losing synchrony with ripple interaction zones.

The paradox collapses when you realize space was never holding anything to begin with. It was merely a relational interface—a perception of distance, not a storage mechanism.

Assumption 2: Information = Matter

The second assumption is even more deceptive: that information is tied to objects, and once the object is gone, so is its information.

This is the root of the paradox:

- A book falls into a black hole.
- It disappears.
- Its particles vanish.
- Its words, story, meaning—gone?

No. That's material reductionism at its most naïve.

UED restores nuance. Information isn't in the object—it's in the field configuration the object once embodied. That ripple pattern isn't erased—it's embedded.

The matter dissolves. The memory echoes.

So when black holes "radiate" seemingly random energy, it's not randomness—it's unresolved ripple discharge.

What we mistook as lost information was simply information realigned beyond our frame of coherence.

Assumption 3: Geometry = Mechanism

Finally, the greatest misstep: believing geometry acts.

Einstein's curved spacetime gave us elegant equations, but also dangerous illusions. When Hawking extended this geometry toward singularities and radiation, it deepened the trap.

In UED:

- Geometry does not cause gravity—it's how we describe ripple tension.
- Horizon boundaries aren't physical limits—they're perceptual thresholds.
- Time doesn't slow or end—ripple response delays, creating the illusion.

Hence, the event horizon isn't a place where physics breaks. It's a zone where misalignment stretches beyond interaction.

Paradoxes only exist when we expect curved geometry to act like a physical machine.

Hawking Radiation: Misunderstood Ripple Leakage

Let's be blunt:

Hawking radiation is not false. It's just misframed.

In UED, the slow "evaporation" of a black hole doesn't signify lost information—it indicates ripple pressure being discharged.

- This discharge appears random if you're blind to ripple

encoding.

- It appears thermal if you only measure amplitude, not phase structure.
- It appears irreversible if you deny non-local resonance.

But when ripple memory and residual echo are acknowledged, this radiation becomes informationally rich, not depleted.

Information "Loss" = Illusion of Shallow Field Models

We didn't lose information.

We lost depth.

Mainstream physics flattened the field into manageable dimensions and boundaries. But reality isn't bound by our simplifications.

- In the full ripple topology of UED, information disperses across non-linear alignments.
- In shallow models, this appears as "loss" only because we stopped listening deep enough.

Hawking's genius was in seeing something strange was happening.

UED's gift is in explaining: it was never strange—it was misperceived.

Closing Statement

The information paradox doesn't need to be solved. It needs to be outgrown.

When space is no longer a vault... When information is no longer a possession...When geometry is no longer mistaken for cause...

The paradox dies—and clarity is born.

"Nothing was lost—only translated. The paradox was never in the black hole. It was in the blindness of the framework that asked the question."

Thought Experiment: The Bell That Never Stops Ringing

Theme: What is gone in form is not gone in field. The essence of information is resonance—not retention.

A Thought Experiment from the Edge

Imagine a bell—an ancient, beautifully tuned bell forged in perfect symmetry.

You walk up to a ripple-active field. It's not a vacuum in the classical sense—it's Structured Vacuum Energy (SVE), alive with invisible tension, balance, and flow. You raise the bell and let it fall into the field.

At the exact moment the bell hits the field's surface, you hear its last tone. A rich, singular *clang*.

Then—nothing. The bell vanishes from sight. No fragments. No remains. No echo—at least to the untrained ear.

But the truth? The sound never stops.

What Happens When You Ring the Universe?

In classical terms, you might say: "It's gone. The object is lost." In relativistic terms: "It crossed the horizon—information inaccessible."

But from UED's point of view, what happened is radically different.

Let's zoom in—not visually, but ontologically:

- The bell created vibrations, ripple alignments that propagated through the SVE.
- These ripples aren't confined to the location where the bell vanished.
- The tone of the bell becomes an encoded ripple signature within the field.
- As it spreads, it interferes, diffuses, layers itself with ambient field tensions.
- Eventually, the signature becomes indistinct—but not *absent*.
- It's *everywhere*—just no longer shaped like a bell.

That's ripple memory.

Misunderstood Loss: It Didn't Go Silent

When mainstream physics says black holes "erase" information, it imagines the bell is gone—and so is its song.

But that assumption is form-centric.

UED reframes this: the form may disappear, but its effect does not. It just changes register.

Just like a bell in an open meadow rings out and the sound fades—but in principle, never completely dies—so too does any form entering a black hole *continue* through the field as distorted wave memory.

This is not fantasy. It's field consequence.

Black Holes as Resonance Distributors

Black holes aren't tombs. They are transformers—zones where information exits structured form and enters resonant dispersion.

- They de-structure.
- They re-encode.
- They re-deploy information—not in ways humans easily interpret, but in a field-level holographic spread.

Imagine every star, atom, or astronaut that crosses a black hole boundary—not as being destroyed, but as being diffused.

Their information doesn't fall into silence. It rings out—through the ripple fabric of the cosmos, indefinitely.

Ask the Reader

What if we've misunderstood black holes because we've confused visibility with existence?

What if:

- To disappear is not to vanish—but to redistribute?
- To lose coherence is not to lose meaning—but to change media?
- To cross an event horizon is not to be erased—but to be scattered in such complex harmony that we no longer recognize the signal?

Then maybe...

Black holes don't end stories—they remix them into the background music of the universe.

And if we could listen *deep enough*, beyond frequency, beyond time— We'd hear the bell still ringing.

"The bell was never lost. Only your ears grew too narrow to follow its echo."

Closing Reflection: Nothing Is Ever Lost — Only Out of Tune

The permanence of information lies not in its visibility or retention—but in its capacity to resonate across ripple fields.

A Return to First Principles

What have we truly lost when something disappears?

In conventional physics, disappearance often implies destruction. The object no longer emits, interacts, or reflects. In the context of Hawking's radiation and black holes, this means irreversible loss—information swallowed, randomness returned, and causality severed.

But UED refuses that verdict.

It doesn't romanticize information—it reframes it. It doesn't argue that things "must be somewhere" in a classical sense. Instead, it states: nothing that ever aligned can ever be unaligned without consequence.

Every object, idea, experience, or field expression creates ripples in SVE. If those ripples become distorted, tangled, or temporally delayed—that's not loss. That's misalignment.

Destruction, then, is simply tension that hasn't resolved. Forgetting is delayed coherence. Memory is resonance that never stopped vibrating—we just stopped tuning in.

UED's Compass: Consequence, Not Collapse

Let us reframe once more.

- Matter collapses in classical theory.

- Information evaporates in relativistic paradox.
- Conservation struggles to explain entropy.

But in UED:

- Matter reorganizes into ripple form.
- Information disperses into field resonance.
- Consequence preserves every change in alignment—whether visible or not.

Even in the most extreme cases—black holes, cosmic annihilations, or quantum decoherence—nothing is lost to oblivion. It merely becomes inaccessible *through conventional interfaces.*

A Universe that Remembers Differently

Your thoughts, your actions, your being—each ripple the field. Even silence makes a mark.

If you've ever wondered where a memory "goes"...If you've felt the lingering echo of something that should've been gone...That's not magic. That's resonance still whispering in a dimension your senses haven't retuned to.

What Comes Next: Radiation and Perception

This closing prepares us for a critical turn.

If information is never lost—but spread, encoded, delayed—then how does it *return*?

- Can radiation reveal truth—not randomness?
- Can perception restore signal—not just observe noise?
- Can we rebuild understanding from field tension?

These questions await us in the next chapter, where Hawking's idea of black hole radiation is reinterpreted not as loss—but as a complex form of echo. Not entropy, but encrypted consequence.

Final Quote:

"The universe does not erase. It rephrases."

It doesn't forget you. It simply waits for the moment when you're tuned well enough to remember what was never gone.

Chapter 5: The Real Meaning of Hawking Radiation

Hawking Radiation is not particle escape from a vacuum edge, but a ripple field's response to accumulated misalignment and energetic fatigue within Structured Vacuum Energy (SVE). Dissipation is a release—not a violation.

Imagine standing at the edge of a black hole—not as an astronaut, but as a physicist trained to believe that nature tears itself apart at such thresholds. You've been taught that pairs of virtual particles spontaneously erupt from the vacuum: one falls in, the other escapes. That this "quantum accident" creates a subtle glow—Hawking radiation—and slowly evaporates the most mysterious object in the universe.

It's a poetic vision. But what if it's just that—a metaphor dressed as mechanics?

What if nothing ever "popped" into existence?

What if no particles leapt free?

What if, instead, what we mistook for radiation was simply the field sighing under pressure?

Structured Vacuum Energy (SVE) doesn't need to invent particles to explain loss. It already carries memory. It already holds tension. And when that tension reaches critical misalignment, it doesn't need drama. It just begins to release.

Not by chance. Not by trickery. But by consequence.

This chapter isn't about denying Hawking's brilliance—it's about liberating his insight from the illusions of geometry and restoring it to the language of ripple logic. Let's listen again—not for what escaped, but for what tried to **realign**. The universe wasn't radiating secrets—it was releasing stress. That's not evaporation.

That's ripple exhaustion.

Revisiting Hawking's Explanation

Stephen Hawking's theory of black hole radiation, introduced in the 1970s, was nothing short of revolutionary. It bridged quantum field theory and general relativity in a way that no one had dared before—and, in doing so, introduced a new kind of paradox that continues to challenge theoretical physics.

Let's revisit his logic—line by line—not to dismiss it, but to understand where its brilliance ends and where its limitations begin.

Quantum Vacuum Fluctuations at the Event Horizon

At the heart of Hawking's explanation is the notion that what we call "empty space" is not truly empty. The quantum vacuum, as predicted by quantum field theory, is a restless sea of activity. Particles and antiparticles are said to *spontaneously* pop in and out of existence, borrowing energy from the vacuum and annihilating each other so quickly that they leave no observable trace.

This constant bubbling of energy is allowed by the Heisenberg uncertainty principle, which tolerates short-term violations of energy conservation—so long as they're brief enough to remain undetected by standard measurement.

Virtual Particle–Antiparticle Pairs

Hawking took this concept and applied it to the region near a black hole's event horizon—the boundary beyond which nothing can escape. His key insight was: what if one of these spontaneously created virtual particles fell into the black hole, while its partner escaped?

In this picture, the escaping particle becomes "real," carrying away energy. The one falling in is absorbed, and—crucially—it carries negative energy (from the point of view of the black hole), effectively reducing the black hole's mass over time.

This is Hawking Radiation: a slow leakage of energy, not from within the black hole itself, but from its very edge.

Time: The Black Hole Shrinks

Over vast timescales, this process results in the black hole gradually losing mass, shrinking in size, and eventually evaporating completely—leaving behind nothing but radiation. Hawking himself described this as the slow death of black holes through quantum means.

But There Are Problems

As elegant as this framework sounds, it raises troubling implications—especially when viewed through the lens of fundamental physics:

- **Randomness**: The particles that escape seem to do so without any encoded trace of the matter that fell in. This radiation is "thermal"—it carries no memory, no message, no correlation to the original input.
- **Energy Imbalance**: The accounting of negative and positive energy in this model rests on highly interpretive

math, not on measurable mechanisms. Assigning "negative energy" to infalling particles is a trick of coordinate systems, not a demonstrated physical reality.

- **Untraceable History**: Most troubling of all—if the information about the original infalling matter is *truly* lost, then quantum unitarity (the bedrock principle that information is conserved) is violated. This breaks the time-symmetry of physics and throws causality into chaos.

Hawking's famous lament captured this unsettling possibility:

"God not only plays dice, but sometimes throws them where they can't be seen."

Thus, while Hawking's model captures the imagination, it also introduced a paradox at the core of physical law. Radiation without memory. Shrinking mass without reconciliation. Information lost in a universe that, until then, was thought to remember everything.

UED doesn't reject this framework out of arrogance—it responds out of necessity. Because there's a deeper mechanism at play—and it doesn't require dice, or disappearance.

It only requires ripples under pressure—and a field intelligent enough to hold the echoes.

UED's Critique of Pair Production

Stephen Hawking's pair production model—a cornerstone of his explanation for black hole radiation—offers a compelling image: particles flickering into existence, splitting at the brink of nothingness, one falling in and one escaping, like a quantum coin toss at the edge of reality.

But in the Universal Energy Dynamics (UED) framework, this picture is not rejected out of contrarian instinct. It is deconstructed with precision. UED sees this model not as a lie, but as a metaphor mistaken for mechanism.

Let's examine what's flawed—philosophically, scientifically, and consequentially.

Particle Pairs: A Visualization, Not a Verifiable Reality

The particle–antiparticle pair near the event horizon was never observed—it was inferred. Born from the language of quantum field theory, it operates in a probabilistic domain where physical intuition often collapses.

UED responds: this isn't how nature works—it's how equations behave when separated from underlying field logic.

What appears as "pair production" is a symbolic construct meant to explain energy transfer near extreme boundaries. But no one has ever caught these "particles" in the act. They are placeholders—an attempt to anchor reality in the familiar form of billiard-ball physics, even when that form doesn't apply.

Vacuum Doesn't Spawn — It Responds

UED rejects the notion of a "fertile vacuum" that randomly gives birth to particles. This idea stems from treating the quantum vacuum as a theater of chaos, where unpredictability is worshiped as insight.

Instead, UED proposes that the Structured Vacuum Energy (SVE) is a coherent, ripple-responsive field. It doesn't spawn—it realigns under pressure.

Radiation, in this view, is not the "birth" of something new. It is the realignment release of stored misalignment. It is consequence, not creation. It is fatigue, not fluctuation.

Energy Doesn't "Borrow" or "Cheat"

The standard explanation allows particles to temporarily borrow energy from the vacuum—as long as they "repay" it quickly. This quantum sleight of hand is permitted by the Heisenberg uncertainty principle, but it has become dangerously misunderstood.

UED draws a hard line here.

Energy is not a bank account that allows overdraft. There is no borrowing, no gambling, no cheating the cosmos.

What we interpret as "borrowing" is actually ripple displacement—an adjustment of field pressure under dynamic stress. The system is always in balance; we just misread the signal when using particle-based language.

The Fatal Bypass: Ignoring Ripple Coherence

The greatest error in the pair-production metaphor is not in what it suggests—but in what it skips.

It bypasses the coherence of ripple fields entirely.

Ripple coherence is the foundation of form, function, and feedback in UED. Every action, distortion, or displacement within the field obeys a law of continuity—no effect occurs without realignment pressure, no emission without pre-aligned tension.

The pair model treats radiation as a spontaneous fracture, not a ripple exhaustion. It lacks the continuity of field interaction. It replaces structural intelligence with statistical improvisation.

Closing Insight: Why This Matters

Hawking's metaphor may have worked in the era of "shut up and calculate" physics. But we now stand in an age where visual stories must meet mechanistic integrity.

UED doesn't reject radiation.

It redefines it—not as particles escaping but as fields exhaling.

"Where others saw particles born from nothing, we see the field whispering under tension—never breaking the law, only revealing its rhythm."

Ripple Realignment Under Pressure

In classical and quantum models, radiation—especially in extreme environments like black holes—is often interpreted as either particle escape (Hawking) or probabilistic tunneling. But UED offers a radically different insight:

Radiation is not emission from particles—it is release from pressure.

Not a loss.

Not an escape.

But a ripple offloading — a necessary realignment when the Structured Vacuum Energy (SVE) field hits its tolerance threshold.

Let's walk through this, not as metaphor, but as mechanism.

Ripple Misalignment as Field Stress

Every ripple is a configuration of tension—structured alignment within SVE. When a mass or energetic event disturbs the ripple field, those ripples compress, skew, or warp.

This distortion is not passive. The field remembers alignment and constantly tries to restore it.

But if the distortion is too intense or prolonged, the system enters a state of field stress — a buildup of uncorrected misalignments.

Think of it like a tightly wound spring. The more you twist, the more pressure builds—not linearly, but exponentially.

Critical Tension and the Need to Release

There is no such thing as infinite storage in SVE. The field can handle misalignment—but only up to a threshold.

As ripple tension approaches criticality, the system does what all natural systems do when overloaded:

It leaks.

But unlike a crack or explosion, this release is subtle, structured, and proportional.

It's not radiation in the traditional sense—it's ripple release.

UED calls this partial realignment under tension. It's the field saying:

"I cannot hold this much distortion anymore, so I'll release a bit—not the whole, just enough to stabilize."

And what do we perceive from this release?

A ripple signature.

A field-level emission.

Radiation—not as light or particles, but as attenuated ripple discharges.

Misinterpreted as Evaporation

In Hawking's view, this is seen as black hole evaporation. Mass converts to radiation. The black hole "shrinks."

But from UED's perspective, nothing truly shrinks or evaporates.

The system simply offloads ripple distortion that exceeded its holding threshold.

What we call "radiation" is a byproduct of misalignment recovery—not of decay, loss, or destruction.

It's a field rebalancing act, not a countdown to disappearance.

The UED Analogy: Thermal Noise Is Delayed Realignment

Take the analogy of thermal noise—what we usually call random vibration or entropy.

Mainstream physics treats this as disorder—a statistical inevitability in a chaotic system.

But UED reframes it as residual misalignment—ripples that haven't yet settled, still seeking their natural harmonic state.

In this view:

- Thermal noise = misaligned ripple tailings
- Heat = ripple density trying to distribute
- Radiation = field tension slowly unwinding

Nothing is wasted. Nothing is random. Everything is ripple behavior under stress.

Conclusion: Radiation Is the Field's Way of Breathing

To an observer tuned into matter, Hawking radiation looks like energy vanishing. To an observer tuned into UED, it looks like energy reconfiguring.

The field doesn't lose—it lets go. Not as sacrifice, but as survival.

"Evaporation is the metaphor. Realignment is the mechanism."

Field Fatigue: A New Lens on Radiation

Mainstream science teaches us that when black holes radiate, they're slowly dying—losing mass by letting particles escape their boundaries. Hawking's radiation model imagines this as an elegant quantum trick: particle pairs dancing on the edge of existence, one falling in, one flying free. But this poetic metaphor hides a deeper flaw: it treats the black hole as a container and radiation as a loss.

Universal Energy Dynamics (UED) invites us to look at this differently.

Not as evaporation, but as relief.

Not as particles being emitted, but as ripples being released.

And at the heart of this new view lies a powerful idea: field fatigue.

What Is Field Fatigue?

Field fatigue is not mechanical wear and tear. It is ripple-level exhaustion.

In UED, the Structured Vacuum Energy (SVE) is not an empty void—it is a highly responsive, ripple-active medium. Every form, force, or function arises from the way ripples align and misalign in this field.

When a system—like a black hole—contains extreme ripple tension, the SVE surrounding it begins to show signs of stress. Not in the material sense, but in its capacity to maintain alignment under prolonged distortion.

This is field fatigue:

A state of sustained ripple congestion, where the field's ability to balance alignment pressures begins to fail—not catastrophically, but gradually and conditionally.

Misalignment Persists — And the Field Responds

Black holes aren't swallowing matter like cosmic vacuums.

They are zones of intense ripple saturation. The misalignment they produce doesn't vanish—it accumulates.

Just as overheated machinery emits occasional sparks, a ripple-saturated field begins to spontaneously discharge microbursts—tiny, often non-linear emissions of ripple energy designed to prevent total collapse.

This is not emission in the classical sense.

It's discharge by necessity—the field's way of whispering "I can't hold this forever."

These discharges are not localized events. They are systemic ripple emissions—carrying the signature of the black hole's internal state, though not its form or content.

Ripple Emissions vs. Particle Escapes

What escapes from the edge of a black hole is not a particle. It's a consequence.

Ripple emissions are not matter leaving—it's stress leaving. A black hole doesn't radiate energy in packets—it relieves pressure by releasing structured ripple echoes, which observers detect as radiation.

Like a steam valve, the black hole doesn't open its contents—it lets off ripple heat from sustained overload.

From the outside, this looks like radiation.

From within UED, this looks like release through resonance.

The Black Hole Doesn't Radiate — It Relieves

This is the most fundamental shift:

Radiation isn't decay—it's adaptation.

Black holes are not decaying stars. They are field saturation nodes undergoing ongoing, necessary ripple relief.

No particles explode. No information is deleted. No magic tricks happen at the event horizon.

Instead, the universe balances itself—quietly, rhythmically, and endlessly—through the logic of ripple fatigue and alignment recovery.

Radiation is the field's sigh. A long, slow breath of recalibration. Not of loss. But of grace.

Signature Leakage: Radiation with Memory

In Hawking's original formulation, black hole radiation is fundamentally thermal—meaning it carries no specific information about the matter that fell into the black hole. The particles emitted, in theory, have random properties, indistinguishable from those emitted by any other black hole of similar mass and spin.

This leads directly to the infamous information paradox: if black holes only emit thermal radiation, then all the information about what entered the black hole is irreversibly lost to the universe. And that, according to quantum mechanics, should never happen.

But here's the twist—

What if the radiation isn't random at all? What if the field is not erasing information, but encoding it—just in a different format?

This is the UED proposition. Welcome to Signature Leakage.

Hawking Radiation = Thermal and Amnesiac

In the traditional view:

- Hawking radiation is blackbody-like: it follows a thermal distribution.
- This implies no distinguishable trace of the black hole's history.
- Radiation is probabilistic—one photon or neutrino is as meaningless as the next.
- This randomness fuels the paradox: if no pattern survives, no past survives.

In essence, heat without history.

UED Radiation = Patterned and Field-Encoded

Under Universal Energy Dynamics, this view collapses. In UED:

- Radiation is not emitted as particles.
- It's a ripple discharge: a structured response to field tension.

- And all ripple activity is inherently patterned—defined by the misalignment it is trying to relieve.

Thus, what escapes a black hole is not a blank emission—it's a signature leakage.

Each ripple carries the tension fingerprint of the system from which it emerged.

Not a snapshot of matter. But an echo of the ripple stress that system endured.

What Is Signature Leakage?

Signature leakage is the field's way of retaining meaning during release.

Here's how it works:

- Ripple radiation is shaped by the precise configuration of misalignments within the black hole.
- These configurations resonate into the Structured Vacuum Energy (SVE) as encoded pulses.
- Just like a voice recorded on vinyl: the vinyl doesn't preserve the voice itself—but the ripples of pressure left behind.
- In UED, the radiation is like that: not the voice, but the traceable pressure.

Each emission, then, is a consequence of what was—not a copy, but a resonant echo.

The field does not need to hold particles. It only needs to remember the pattern of pressure they left.

Not Information Loss — But Encoded Resonance

Here is the most crucial insight:

Radiation in UED is not forgetting—it's reframing.

- What seems like lost information in GR is merely unrecognizable due to our particle-centric perspective.
- The information is not destroyed, but dispersed—in non-local, non-linear, holographic ripple form.
- From a thermal lens, it looks noisy.
- From a ripple lens, it's symphonic—if you know how to listen.

In other words:

- Hawking's model assumes loss because it looks for repetition.
- UED assumes conservation because it honors resonance.

Final Analogy

Imagine this:

You strike a unique bell. The sound spreads in all directions. Later, you listen to distant echoes and say, "This is just noise."

But to a trained ear? That noise carries everything—the size of the bell, the force of the strike, the angle of the swing.

The same applies here. What Hawking saw as thermal fog, UED sees as echo-rich air.

The Shrinking Illusion

Mainstream physics tells us that black holes shrink.

Bit by bit, Hawking radiation drains their mass—until, supposedly, they evaporate completely, vanishing in a final, ghostly burst of energy.

But that's only half a story.

From the lens of Universal Energy Dynamics (UED), this so-called evaporation isn't an escape of mass—it's the unraveling of misalignment.

What appears to be shrinkage... is actually realignment.

Black Holes Don't Evaporate — They Offload Misalignment

Let's revisit what a black hole *is* in UED:

Not a singularity. Not a spatial sink. But a ripple tension zone—a region where Structured Vacuum Energy (SVE) has reached peak misalignment saturation.

What builds pressure inside a black hole is not density alone, but the inability of ripples to reconfigure harmoniously.

Eventually, this strain forces the system to leak alignment attempts—ripple discharges that we perceive as radiation.

This is not matter *leaving*. This is tension easing.

So what's really shrinking?

Not the existence. Not the essence. Just the misalignment index.

Structural Reorganization, Not Disappearance

Under Hawking's framework, black hole mass decreases over time due to emitted radiation. Under UED, what's happening is more nuanced:

- Matter within the black hole does not *evaporate* into non-being.
- It dissolves into ripple form—a transition from massive structural coherence to distributed ripple resonance.
- The *form* changes. The *consequence* remains.

This shift doesn't destroy structure—it redistributes it across the field.

What looked like loss was simply a change of format.

From solidity to signature. From storage to broadcast.

Energy Is Never Lost — It Shifts Form

UED holds firmly to a foundational principle:

Nothing in the universe disappears. Everything transforms.

Let's trace the ripple logic of transformation:

From Matter to Ripple

- Dense matter under extreme ripple tension reaches a point where structural coherence can no longer hold.
- It dissolves—not into particles, but into field-aligned waveforms.
- These ripple expressions are no longer *contained*—they become field-active consequences.

From Storage to Expression

- A black hole is a repository of misalignment.
- As ripple fatigue sets in, it begins to express what it can no longer hold.

- This is not escape—it is release.
- The system isn't "losing mass." It's converting tension into communication.

In this way, the field returns to balance not by rejecting energy, but by rephrasing its stress.

The Illusion of Shrinking

When we say a black hole "shrinks," we imply that it's disappearing.

But in UED terms:

- The visible mass shrinks because ripple tension reduces.
- The energy isn't gone—it's simply no longer in the form we associate with mass.
- What we observe as shrinkage is really the field exhaling.

It's not a death. It's a decompression.

Final Reflection

What if black hole evaporation isn't about loss—but about return?

Return of alignment. Return of balance. Return of rhythm to a part of the field that once overflowed with strain.

Shrinking is not surrender—it's the system choosing peace.

Consequence of Continuous Realignment

In mainstream thinking, radiation from a black hole is seen as spontaneous emission—a sort of random quantum gamble.

In Universal Energy Dynamics (UED), nothing is spontaneous. Not because chance doesn't exist, but because even chance must consequently emerge from imbalance.

There is no emission without tension. There is no release without pressure. There is no radiation... without a reason.

Radiation Is Not Spontaneous — It's Consequential

Let's correct a foundational error in current models:

- Hawking radiation, as theorized, emerges from vacuum fluctuation.
- These fluctuations are treated as random, probabilistic anomalies at the edge of the event horizon.
- But randomness is not a mechanism—it's a label we use when we don't understand the field conditions.

UED reframes the phenomenon:

Radiation is not a surprise event. It is the inevitable response of the Structured Vacuum Energy (SVE) field under continuous ripple stress.

Just as a fault line builds pressure and eventually causes an earthquake, Just as a machine overheats and vents steam, The ripple field, under distortion, must discharge alignment attempts to avoid catastrophic incoherence.

This is not a flare. This is rebalancing.

The Greater the Misalignment, the Greater the Urge to Dissipate

In UED, ripple fields possess intrinsic harmonic intelligence—a natural drive toward equilibrium.

When parts of the field fall out of tune due to:

- Intense gravitational accumulation
- Chaotic interactions
- Collapse into ripple-dense zones (like black holes)

...the field begins to resist the distortion.

But this resistance is not passive. It stores. It tightens. It compacts. Until the only remaining option is release.

Not escape. Not destruction. But realignment under duress.

The longer realignment is delayed, the stronger its eventual expression.

So what we call radiation is the field's sigh, after holding its breath for too long.

No Black Hole "Dies" — It Evolves

Let us remove another myth: the *death* of a black hole.

Mainstream physics imagines:

- A black hole loses mass via Hawking radiation.
- Eventually, it shrinks to nothing.
- It "dies."

UED says otherwise:

- Black holes do not die—they transform.
- The misalignment that defines them slowly dissipates across field-time.
- Their identity as a ripple congestion zone dissolves.
- But their essence? It's not erased.

That essence—what they once held, twisted, suppressed—is reformulated as field resonance.

Nothing ends. It reintegrates.

Field Coherence Is Reclaimed Gradually

Here lies the beauty of ripple logic:

- Even the most misaligned zone has a path back to harmony.
- But it cannot be sudden—it must be progressive, as the system allows.
- Black hole "radiation" is not explosive—it's convalescent.

As the ripple field discharges:

- Echoes fade.
- Tension drops.
- Coherence returns.

This doesn't happen on our clocks. It happens on alignment time—the rhythm of field reconciliation.

Radiation is not just the end of a cycle. It's the slow reweaving of the fabric that was once too tight to breathe.

Final Thought

What if black holes aren't black at all?

What if they're darkened only by our current inability to read their realignments?

UED suggests: they are not voids, but repositories of potential, held in suspension by the very intelligence of the field.

Their radiation is not the gasp of a dying object... ...it's the whisper of a system finding its voice again.

Rethinking Black Hole Lifecycles

We've been trained to imagine black holes like ticking time bombs—ominous voids with a finite countdown.

Stephen Hawking famously predicted that black holes lose mass through radiation, and with enough time, they vanish.

This idea—known as black hole evaporation—suggests a definitive end. A fade-out. A death.

But Universal Energy Dynamics (UED) doesn't see the cosmos as something that dies.

It sees it as something that transforms.

Hawking: The Vanishing Act

According to Hawking's framework:

- Black holes radiate via quantum vacuum fluctuations at their event horizons.
- The process is extremely slow, but over billions or trillions of years, a black hole is expected to:
    - Shrink in mass,
    - Intensify in temperature,
    - And ultimately vanish in a burst of final radiation.

What remains? According to this view: Nothing. Just a statistical trail of thermal randomness. The cosmic equivalent of an erased file.

But that assumption depends entirely on two flawed premises:

1. That information is lost when structure is lost.
2. That "nothingness" can be the result of transformation.

UED: No Vanishing, Only Realignment

In UED, the black hole lifecycle is not a slow death—it's a slow rebalancing.

Let's consider how:

- A black hole, from the UED standpoint, is a ripple tension node—a location in the Structured Vacuum Energy (SVE) field where misalignment, distortion, and pressure have reached critical saturation.
- Over time, the field attempts to offload this pressure through ripple emissions (what conventional physics calls Hawking radiation).
- But this isn't "evaporation" in the thermodynamic sense.
- It is field reabsorption—the system digesting its stress.

So, what's left behind isn't an empty canvas—it's a softened ripple zone, one where:

- The original field tension has eased,
- The misalignment has been partly realigned,
- The chaos has been converted into subtle, new resonance.

The Afterlife of a Black Hole: Echoes in the Field

Here's the real shift: A black hole doesn't disappear—it changes identity.

What was once:

- A dense knot of incoherent ripple activity,

- Resistant to perceptual alignment,
- Opaque to information exchange,

...becomes:

- A calmer node of residual ripple flow,
- A region of complex but less hostile field geometry,
- A new signature in the ripple map of the universe.

It may no longer be "detectable" by the same gravitational lensing or radiation models, but it leaves behind a traceable fingerprint—not unlike a scar in spacetime, except in UED, it's not scar tissue. It's adaptive tissue.

Not an Absence — A Signature

Let's reframe the outcome.

What if the end of a black hole was not:

- A hole closing,
- A thing disappearing,
- An object being destroyed,

...but rather:

- A field completing its cycle,
- A system shedding its load,
- A new waveform emerging from exhausted turbulence?

This is what UED refers to as a ripple field signature—a lasting but altered pattern in the SVE topology that holds memory, not mass.

You may no longer see the monster, but you can still hear its echo.

You may no longer feel the force, but you'll still notice the bend in the breeze where it once stood.

Beyond Finality: Toward Evolutionary Cosmology

Traditional physics often leans on closure metaphors: collapse, death, vanishing.

UED replaces them with continuity metaphors: reconfiguration, resonance, conservation.

So rather than asking:

- *"When does a black hole end?"*

We ask:

- *"What does it become next?"*

Because in a universe governed by consequence—not annihilation—nothing is lost. It just changes what it offers back to the field.

The death of a black hole is not a vanishing point. It is a consequence field, folded back into harmony.

Closing Reflection: Radiation Was Never a Betrayal

We once feared that black holes were traitors to the laws of nature.

That they pulled in order and returned chaos.

That they took in storylines—stars, planets, even memories—and spat out only meaningless noise.

We called that noise "Hawking Radiation."

We framed it as a betrayal of the cosmos. As if the universe had allowed a loophole where cause vanished and consequence failed.

But what if that wasn't the case at all?

Radiation Wasn't Random — It Was Rhythm

From the perspective of Universal Energy Dynamics (UED), what radiates from a black hole isn't a random discharge, but a resonant offloading—the field's intelligent response to internal stress.

What seems like:

- Thermal noise in classical theory
- Quantum evaporation in Hawking's model

...is in fact a signature discharge—the ripple field releasing exactly what it can, when alignment permits, and as much as the surrounding SVE can accommodate without destabilization.

Radiation, therefore, isn't a breakdown. It's a breath. A field exhaling tension.

Black Holes Aren't Leaky Graves — They're Consequence Engines

The metaphor of black holes as "cosmic cemeteries" is emotionally compelling but physically misleading.

UED reframes them as temporary consequence zones—regions where field pressure builds up, reconfigures, and slowly releases its tension back into the universal lattice.

In this view:

- They don't erase.

- They don't cheat.
- They don't destroy meaning.

They process it.

Radiation isn't the field giving up on what was inside. It's the field finding a way to re-express what couldn't stay structured.

Just because the form is gone doesn't mean the story is.

The Field Always Knows What It's Releasing

One of the most elegant aspects of UED is its assertion that the field is never blind.

It doesn't operate on probabilities or ignorance. It operates on consequence alignment.

What appears as randomness in Hawking's model is actually deferred patterning in UED.

Just as steam from boiling water carries the thermal memory of its source, ripple radiation carries encoded traces of the field structures it emerged from—even when we can't immediately decipher them.

So the field doesn't gamble. It recalibrates. It doesn't erase its past. It translates it.

A Universe Without Dice

Stephen Hawking once famously declared:

"God not only plays dice, but sometimes throws them where they can't be seen."

UED replies with a gentle correction:

"The universe never throws dice. It merely redirects them with rhythm."

What looks like chance is just consequence too complex to compute.

What looks like chaos is structure unrecognized by our current models.

Rephrasing Radiation

So let's reframe what Hawking radiation truly is:

- It's not leakage. It's alignment leakage.
- It's not loss. It's latent resonance.
- It's not betrayal. It's misunderstood intelligence in motion.

Hawking wasn't wrong to see that something radiates. But perhaps the radiation wasn't a signal of loss—it was a ripple still speaking.

A ripple we had not yet learned to listen to.

"Hawking's radiation wasn't a mistake. It was a misread ripple—trying to speak, not disappear."

Chapter 6: Entropy, Alignment, and the Arrow of Misunderstanding

Entropy is not chaos, decay, or heat death — it is the measure of ripple misalignment in Structured Vacuum Energy (SVE). The so-called "arrow of time" is not driven by disorder, but by the field's response to ripple distortion and its struggle to realign. UED reframes entropy as a consequence of incoherence, not a law of inevitable decline.

Picture yourself in a quiet room. A cup of freshly brewed tea sits on the table, steam rising gently into the air. Over time, that steam fades. The warmth dissolves. The tea settles to room temperature. Classical physics says: "Entropy has increased." The system has lost order, drifted toward randomness, and can no longer do useful work.

This, we are told, is nature's law: that energy disperses, order decays, and everything—eventually—fades into thermal sameness. Entropy becomes the story of everything winding down. The ultimate accountant of the universe's decline.

But let's look again.

What really happened when the tea cooled?

Did the universe become more chaotic? Did some invisible god of disorder press a button labeled "Entropy++"? Or did the tea simply come back into balance with the field around it?

Under the lens of Universal Energy Dynamics (UED), the story is different. The steam isn't vanishing into chaos—it's dispersing into equilibrium. The warmth isn't being lost—it's being redistributed into the field. The molecules aren't rebelling—they're relaxing. The system isn't collapsing into disorder—it's realigning with its background structure: the Structured Vacuum Energy (SVE).

The tea didn't die. It found harmony.

This is the crack in the second law's armor. Entropy, when viewed without geometry and beyond statistics, isn't the villain we've made it out to be. It's not a blind force of destruction, but the signature of tension being released. It's the language of misalignment resolving itself.

Entropy doesn't mean chaos. It means consequence.

So here's the deeper question: **Is entropy truly about disorder, or is it about the journey back to field coherence?** Is the so-called "arrow of time" pointing toward decay—or pointing us back toward resonance?

Let's pull on that thread. Because once we reframe entropy as ripple misalignment—and not thermodynamic doom—the universe doesn't look like it's dying. It looks like it's tuning itself.

Classical Entropy: From Thermodynamics to Time's Arrow

To understand what UED redefines, we must first revisit what classical physics assumed.

The Traditional View

Entropy, as defined in classical thermodynamics, is a measure of disorder or randomness in a system. The concept, formalized by Ludwig Boltzmann, emerged from statistical mechanics—a framework that interprets macroscopic phenomena based on the behavior of microscopic particles.

At its core, entropy S in Boltzmann's famous equation, $S=k\cdot\ln\Omega S$, quantifies the number of possible microscopic arrangements (Ω) that correspond to a macroscopic state. The more ways the system's particles can be arranged without changing the observable outcome, the higher the entropy.

From this, came a ruling principle of physics: The Second Law of Thermodynamics.

"In a closed system, entropy always increases."

Translated: isolated systems tend to evolve from ordered states (low entropy) to disordered ones (high entropy). This is why hot tea cools, stars burn out, and energy gradients flatten. The law is said to be unbreakable, with entropy acting as nature's slow countdown to equilibrium—or, as some would put it, eventual heat death.

This progression gave rise to one of physics' most mysterious constructs: the arrow of time. While Newton's laws are reversible, and even quantum mechanics can run forwards or backwards, entropy breaks the symmetry. It moves only in one direction—forward. We remember the past but not the future. We age but never de-age. All because entropy marches on.

But there's a problem.

The Hidden Weakness

Despite its predictive power, the classical interpretation of entropy is descriptive—not causal. It tells us what happens, but not why it happens.

- Why must entropy increase?
- Why does the arrow of time emerge at all?
- Why can't we observe the reverse—spontaneous

reordering?

The equations of thermodynamics do not answer these. They summarize trends over billions of interactions but do not define a mechanism for misalignment, decay, or irreversibility. They assume a background geometry—space and time—as neutral containers. They measure outcomes, not field conditions. And most importantly, they ignore the medium through which all energy flows: the vacuum field.

Classical entropy also relies heavily on probability, which, while mathematically elegant, is physically agnostic. It doesn't know what a ripple is. It doesn't track tension in a structured field. It doesn't consider the memory embedded in vacuum fluctuations or the feedback loops of misaligned states. It assumes disorder is the end of the story.

UED challenges that narrative.

What if entropy isn't about disorder at all?

What if it's simply a ripple field reporting its misalignment—a signal, not a sentence?

This opens the door to a radical shift: entropy not as the driver of time, but as the *consequence of ripple delay and structural incoherence* within the universe's most fundamental medium—Structured Vacuum Energy (SVE).

In the next section, we'll dive into that exact redefinition.

The Problem: Equating Entropy with Irreversible Decay

Entropy has become one of the most misunderstood and misused terms in science and pop culture alike. It's often casually associated with chaos, breakdown, and even the inevitability of death. But this association—while symbolically powerful—is fundamentally misleading. In fact, it has created conceptual blind spots in physics that UED directly addresses.

The Cultural Myths of Entropy

Ask the average science enthusiast, and you'll likely hear that:

- Entropy means things fall apart.
- Entropy is the arrow of decay.
- Entropy ensures that no matter what we do, order will always deteriorate into chaos.
- Entropy is why time moves forward, why memories fade, and why nothing lasts.

These statements sound poetic—but they're metaphors, not mechanisms. And like all metaphors, they can obscure as much as they reveal.

This mythologized version of entropy paints it as an unstoppable force of destruction—the villain of thermodynamic destiny.

But here's the catch:

Entropy is not destruction. It is not even disorder. It is a mathematical consequence of how we count possibilities under an assumed structure.

What's Actually Being Measured?

In traditional thermodynamics, entropy is a measure of how many microscopic configurations can produce the same macroscopic outcome. When more configurations are available, entropy is said to be higher. But this assumes we're counting based on particle arrangement—which itself depends on a model of matter as isolated, billiard-ball-like components in space.

This model ignores the deeper reality: fields, coherence, and interaction memory. It doesn't consider whether these microstates are dynamically stable, ripple-aligned, or energetically harmonious. It just counts.

And it does so under the presumption of randomness—not structural tension, not resonance, not ripple harmony.

Why "Decay" Is the Wrong Word

Classical interpretations equate entropy with irreversibility. Once disorder sets in, we assume it's final. But irreversible for whom?

In reality, what we call "decay" is often just a reconfiguration from one kind of coherence to another. A star doesn't decay—it transforms. A cooled cup of tea doesn't die—it rebalances. Even death itself, biologically, is a ripple redistribution of energy, not an annihilation of pattern.

In each case, the pattern is not destroyed—it is disassembled, diffused, or shifted into non-observable alignments. The field still holds the echoes.

This points to a major oversight:

Thermodynamic abstraction treats energy flow as a scalar loss, not a vector realignment.

Entropy becomes a number detached from field intelligence, from vacuum structure, from ripple behavior. It tells us that something has changed—but not how, where, or why the coherence was lost.

It assumes a neutral background, and it calls what it cannot track "irreversible."

But UED argues that nothing is truly irreversible. Everything is consequential, and all consequences are field-accounted.

What Thermodynamics Ignores

Thermodynamic entropy does not:

- Account for ripple misalignment within structured vacuum energy.
- Recognize that so-called disorder is often unmeasurable complexity—not chaos.
- Include the resonance tension or field saturation that actually governs whether systems evolve, transform, or stabilize.
- Incorporate the possibility that information is preserved non-locally, even when physical forms dissolve.

In other words, it misses the field. And if you miss the field, you miss the foundation of how the universe retains, reforms, and re-expresses coherence.

Entropy, then, is not the villain. It is not the reaper of form or the destroyer of order. It is the metric of misalignment within the ever-consequential ripple field that surrounds us.

It's time to replace the metaphor with the mechanism.

UED Reframe: Entropy = Misalignment

In the Universal Energy Dynamics (UED) framework, entropy is not a march toward chaos. It's not an irreversible slide into decay. And it certainly isn't the universe's way of forgetting. Instead, UED makes a sharp departure from conventional thermodynamics by redefining entropy as a state of ripple misalignment within the Structured Vacuum Energy (SVE) field.

From Randomness to Ripple Incoherence

The classical view treats entropy as a statistical measure of disorder—a shorthand for how unpredictable or disordered a system is. But UED asks a deeper question: "Disordered with respect to what?"

The answer lies not in particle positions, but in ripple alignment—the ability of the field to support, sustain, and reconfigure coherent structures.

In UED, entropy is not about unpredictability. It's about the field's internal tension—its resistance to harmonious alignment.

A system is not "high entropy" because it is messy or complex. It is high entropy because its ripple patterns are incoherent, turbulent, or fragmented. These misaligned ripples make the field less capable of expressing organized phenomena like motion, radiation, memory, or even perception.

Ripple Patterns as the New Metric

Let's break it down:

- **High Entropy** → Ripple tension is high. The local SVE is saturated, congested, and unable to align efficiently. Information is buried under misalignment. Systems appear chaotic because the field can't stabilize a pattern.
- **Low Entropy** → Ripple harmony dominates. The field flows cleanly. Energy transitions are smooth. Perception, motion, radiation, and memory express themselves with minimal

resistance.

This is not randomness vs. order.

This is resistance vs. resonance.

Coherence as a Physical Capability

In this model, entropy is not just a description—it becomes a dynamic property of the field's responsiveness.

- Can the field support a ripple structure with minimal loss?
- Can energy enter and exit without creating chaos?
- Can information reform in recognizable patterns?

If the answer is yes, entropy is low. If no, entropy is high—not because something is "disordered," but because the field is inflexible, misaligned, or oversaturated.

This leads to a powerful conceptual shift:

Entropy is no longer a number describing fate. It's a relational state describing readiness.

It tells us not where a system is going, but how well it can realign with universal harmony.

A Consequence, Not a Doom

The Second Law of Thermodynamics often casts entropy as a one-way street—a doom-bound decline. UED dismantles that.

Entropy, in UED, is not directional by default.

- Systems don't always fall into disorder.
- They reconfigure.

- They can recover.
- They can realign.

Entropy increases when misalignment grows—not because the universe demands it, but because field tension is left unresolved.

Entropy decreases when alignment is restored—when coherence is nurtured, resonance achieved, and the ripple field regains balance.

It's not a law of death. It's a measurement of harmony lost—or found.

The Field Is the Framework

The structured vacuum is not passive. It's not an empty stage. It is the carrier of all energy, information, and coherence. And entropy, through the UED lens, is simply its condition:

- How strained is the field?
- How distorted are the ripples?
- How ready is it to reorganize itself?

This makes entropy a field intelligence metric—a feedback on the health, tension, and responsiveness of SVE at any point in space-time.

Conclusion: Entropy Isn't Fate—It's Feedback

UED's reframing doesn't just fix the misunderstandings—it opens up new dimensions of thought.

- It explains entropy without invoking statistical chaos.
- It makes it measurable in terms of field tension.
- And most importantly, it restores agency, recovery, and reversibility to the system.

Entropy is not a ticking clock to oblivion. It's a signal—a whisper from the ripple field saying: "I'm misaligned. Help me realign."

Coherence vs. Disorder: A Matter of Field Tension

In classical physics, the word "disorder" has long dominated entropy discussions. It paints a picture of systems falling apart—atoms scattering, structures dissolving, and the arrow of time dragging everything toward decay. But UED invites a more nuanced—and more accurate—perspective:

What we call disorder is not the absence of order. It is the presence of unresolved tension within the ripple field.

Let's reframe entropy through the lens of coherence—not just in form, but in ripple structure and field alignment.

Misalignment Isn't Anarchy

When ripple coherence is lost in Structured Vacuum Energy (SVE), several things begin to occur:

- **Delayed Responses**: Energy no longer flows smoothly. Motion becomes sluggish. Events that require field agreement—like perception, radiation, memory—get delayed or disrupted. This isn't chaos. It's inertia caused by internal disagreement in the field.
- **Loss of Structure**: Systems lose their ability to sustain clean ripple patterns. Like a soundwave in water muddied by interference, signals begin to dissolve into interference, not disappear. The system doesn't disintegrate—it detunes.
- **Perceptual Fragmentation**: Information becomes harder to extract. This manifests as:
    - Memory loss in cognitive systems.

    - Signal noise in machines.
    - Decoherence in quantum systems.

Again, this is not disorder—it is the field becoming opaque to its own harmony.

UED emphasizes: entropy is not about the breakdown of reality. It's about breakdowns in responsiveness, which occur when the ripple field is too tense to reconfigure clearly.

Healing Through Alignment, Not Time Reversal

The classical idea of reversing entropy to restore a system leads to the myth of time reversal—as if rewinding a tape would put the pieces back together. UED counters this with a more biologically and physically grounded model:

Restoration does not require reversing time—it requires reducing ripple tension.

In UED:

- Healing = reduction in ripple interference.
- Recovery = re-emergence of ripple coherence.
- Reversibility = not going back in time, but rebuilding alignment in present time.

This applies to everything from:

- Neural trauma (restoring memory coherence),
- Radiation damage (re-stabilizing field structures),
- Cosmic evolution (realigning ripples after gravitational stress).

Time doesn't "go backward." The field regains its clarity.

Beware the Word "Disorder"

When we call something "disordered," we often imply that it is *inherently* beyond repair. UED cautions against this fatalism.

- A glass shattering isn't disorder—it's redistributed tension.
- A galaxy collapsing isn't chaos—it's an attempt at reconfiguration under immense field stress.
- A mind forgetting isn't destruction—it's overload or misalignment in cognitive ripples.

What we casually label "disorder" is simply complex coherence failure—and that can be healed.

Summary: Coherence Is a Function of Ripple Readiness

Let's be clear:

- Disorder is not an ontological state. The universe does not default to chaos.
- It defaults to balance—and tension disrupts that balance.
- Coherence is not imposed by external order—it is allowed by internal alignment.

The moment we stop treating entropy as inevitable loss and start seeing it as a tension report from the ripple field, we gain:

- Better scientific models,
- More accurate diagnostics,
- And most importantly, a hopeful framework for everything from black hole dynamics to human healing.

Entropy doesn't destroy. It resists. And resistance can be resolved.

Time's Arrow Misinterpreted

The "arrow of time" is one of the most sacred metaphors in classical physics. From the cooling of tea to the collapse of stars, everything, we are told, marches forward—from past to future, from low entropy to high. But this forward-pointing arrow is not an inherent law of nature—it's a human interpretation based on limited perception.

Universal Energy Dynamics (UED) offers a bold revision:

Time is not a river flowing one way. It is a ripple alignment delay we experience as directional.

Let's break this down.

Classical View: From Low to High Entropy

In conventional thermodynamics:

- Entropy increases over time.
- Ordered systems become disordered.
- Events are irreversible.
- This irreversible increase is what supposedly gives us the "arrow of time."

This idea is deeply embedded in everything from cosmology to psychology:

- Eggs break, but don't un-break.
- People age, but don't de-age.
- The universe expands and cools, never rewinding.

But here's the issue: none of this actually defines time itself—only how we experience change.

UED View: Time = Alignment Delay

UED proposes a shift in framing:

- Time is not a "thing" that flows.
- It is a perceived delay in ripple realignment within Structured Vacuum Energy (SVE).
- In other words, we don't move through time. We move through states of field misalignment, and we *call* that sequence "time."

Here's an analogy:

- Think of walking through fog of varying thickness.
- Sometimes your steps feel fast, sometimes slow—based not on time itself, but on the resistance from the fog.

In UED, this fog is ripple misalignment:

- Where the field aligns smoothly, we experience flow, clarity, ease.
- Where the field is tense or saturated, we experience drag, inertia, confusion—or what feels like "slowness" or "past."

Thus, time is not a line. It is a map of field tension.

The Arrow Is Not Fixed

From UED's perspective, the arrow of time is not a universal direction but a vector of field intelligibility.

Let's decode that:

- Intelligibility means how readable, resonant, and coherent the ripple field is at any point.

- When ripple fields are becoming more misaligned, we interpret this as "entropy increasing" and call it *forward*.
- But in UED, this is simply a vector—a direction of loss or gain in resonance—not a rule of one-way flow.

This means:

- In zones of re-alignment (e.g., healing, reintegration, coherence), entropy decreases—and yet time still flows *for us*.
- Time, then, is not about direction, but about the delay experienced between ripple cause and ripple effect.

Perception and the Illusion of Forward Motion

Why, then, do we *feel* like time only moves forward?

Because:

- We are observers embedded in ripple fields.
- We perceive reality only through the realignment of those fields.
- When ripple coherence improves, we perceive "healing" or "clarity".
- When misalignment accumulates, we perceive "decay" or "aging".

In both cases, we track change, not time. But our brains mistake that change as time's passage.

Think of this like scrolling through a photo album:

- The photos don't move.
- Your finger does.

- Yet you feel like the story is "moving forward."

That's perceptual inertia—not time.

The Cost of Misinterpretation

Treating entropy as the arrow of time has led to several flawed assumptions:

- That the future is inherently worse (due to disorder).
- That systems cannot regain order without external force.
- That death, decay, and collapse are irreversible destinations.

But if time is ripple delay, not cosmic doom, then:

- Reorganization is possible.
- Recovery is not anti-time—it's alignment.
- Evolution isn't a forward march—it's the field's dynamic realignment.

UED reframes time as a measure of struggle between tension and coherence, not a ticking clock dragging us to oblivion.

Summary Thought

Time doesn't move. We do—through the densities of the ripple field.

What classical physics labeled as "irreversibility," UED sees as difficulty in restoring ripple harmony. And what we called the "arrow of time," might just be the direction our perception favors when field alignment degrades.

Entropy and Perception: Observer Matters

Entropy has long been treated as an objective property of systems. In classical physics, it's a statistical measure — how many microstates are accessible, how "spread out" energy is, how likely a system is to move toward disorder. But this statistical view omits one critical dimension:

The observer.

In Universal Energy Dynamics (UED), entropy is not merely a system's property — it is also a relational consequence between the field and the observer's ripple coherence.

Let's explore.

Perceived Entropy Is Relative to Alignment

Imagine two people walking into the same room filled with scattered papers, buzzing devices, overlapping conversations, and flashing lights.

- One says, "This place is a mess!"
- The other — perhaps a conductor, an engineer, or an artist in their element — says, "Ah, perfect! Everything's in motion!"

Same environment. Different perception.

Why?

Because perception is not just sensory intake — it is decoding. And decoding depends on how well an observer's internal ripple pattern aligns with the external field fluctuations.

In UED terms:

- Entropy = misalignment.

- But perceived entropy = failure of the observer to make sense of the ripple state.

So, what seems "chaotic" or "noisy" to one observer may appear structured, meaningful, even beautiful to another — not because the system has changed, but because the observer's internal field is better tuned.

Perception = Ripple Decoding Capacity

Every observer has a bandwidth — a limit to how much ripple distortion they can interpret before it becomes noise.

When field tension exceeds your decoding capacity:

- Signals become static.
- Events lose causality.
- Experience feels fragmented.
- You call it "confusing" or "overwhelming."

But with higher alignment:

- You discern signal from noise.
- You read deeper into complex systems.
- You see patterns others cannot.

Thus, perception is not just about data—it's about resonance with the field.

This is why a quantum physicist finds elegance in probabilistic functions where a layperson sees uncertainty. Or why a poet sees symmetry in emotion where a critic sees inconsistency.

The field didn't change. The observer did.

Alignment ≠ Intelligence Alone

Important distinction: UED does not equate better perception with higher intellect or knowledge.

You can be intelligent and still misaligned.

Ripple decoding capacity depends on:

- Clarity (how free you are from internal turbulence).
- Readiness (your intent and attention).
- Contextual resonance (alignment with what you're observing).

Even a child can perceive high-order coherence in nature when attuned — and even a scholar can miss the point when over-fragmented by abstraction.

That's why:

"To the aligned, even entropy looks like art."

Consciousness: The Observer's Bridge

This is where consciousness enters the UED frame.

- It is not a product of the brain, but a field interface — a way of sensing, aligning, and decoding ripple interaction.
- Consciousness isn't just self-awareness; it's alignment-awareness.

Thus, perception becomes not just observation, but interaction.

- You don't merely watch entropy happen.
- Your awareness collapses ambiguity into meaning — or fails to.

So while a system may be "high entropy" in ripple terms, your experience of it is determined by your conscious field coherence.

Entropy Is Not Objective — It's Relational

This changes everything.

In UED, entropy is no longer:

- A universal death march toward disorder.
- An impersonal tally of lost order.

Instead, it is:

- A ripple mismatch between the observer and the environment.
- A feedback signal about how aligned or out-of-tune we are with what we're experiencing.

This means:

- Entropy has no absolute value — only contextual resonance.
- The universe does not become more chaotic over time — we lose resonance with it, and *interpret* that as chaos.

Final Thought

"Entropy doesn't tell you what the world is. It tells you what you can no longer read."

In this light, entropy is not a fate — it's a message. Not a measure of disorder, but a reminder to realign.

So, the next time the world feels too loud, too fast, too broken—don't ask, "Why is everything falling apart?" Ask instead: "Where have I fallen out of resonance?"

Reversible Universe: If Alignment Is Possible

What if the universe isn't fated to decay into heat death?

What if the so-called "irreversible end" is only the slow rhythm of misaligned ripples struggling to find their beat again?

In classical thermodynamics, entropy dictates a one-way path:

- From order to disorder.
- From energy potential to dissipation.
- From usable form to useless spread.

This linear thinking leads to the grim conclusion of a dying universe — a vast, cold expanse of irretrievable chaos.

But UED offers a radical reframe:

If entropy is ripple misalignment—not permanent disorder—then realignment remains possible.

Entropy Isn't a Wall — It's a Wait

Traditional entropy models assume irreversibility because they mistake delay for finality.

But in the UED framework:

- Entropy = ripple incoherence, not annihilation.
- Misaligned ripples don't vanish — they persist, often tangled, dispersed, or echoing in low-resonance zones.
- The field remains intact, carrying all the information in

distorted forms.

So what does this mean?

- A cup of tea can't reheat itself spontaneously — because the ripple environment no longer supports that realignment without external intervention.
- But theoretically, if the entire surrounding field reconfigured itself to match the initial ripple pattern — the warmth, motion, and memory could return.

This isn't magic. It's field potential.

The question isn't "Is it reversible?" It's "Can alignment conditions be restored?"

The Universe Doesn't Die — It Cycles

UED rejects the notion of a final, irreversible end.

Why?

Because ripple fields are not linear systems. They are dynamic, recursive, and intelligent.

- Field tension builds → releases → stabilizes → reconfigures.
- Misalignment increases → until coherence is restored through consequence (LOC).
- No ripple is ever fully lost — just delayed in rediscovery.

So the "death" of stars, the "collapse" of black holes, even the "cooling" of the cosmos is not destruction—it's a phase of storage, awaiting field resonance for re-expression.

The universe doesn't burn out. It exhales and inhales in field-based rhythms.

Even Black Holes Don't End Information

Under General Relativity, black holes seem like cosmic deletion zones.

But UED reframes black holes as:

- Ripple congestion sites.
- Tension memory banks.
- Zones where misalignment stalls observability, not existence.

Information isn't destroyed in a black hole. It's encoded in ripple distortion — compressed so deeply it can't be read, but still exists within the SVE (Structured Vacuum Energy) topology.

And just like a note too faint to hear isn't silence—

A field too dense to decode isn't void.

When field pressure is released, or new coherence conditions emerge — even black hole-held patterns can begin to reform, radiate, or resonate again.

Reversibility Isn't Resetting — It's Re-Resonating

Let's be clear:

UED does not propose that everything will rewind like a film reel. That's a Newtonian fantasy.

Instead, it asserts something more nuanced:

- No pattern is ever fully lost.

- If enough of the original field context returns, even the most entropic expression can be retuned.

This is why memory sometimes returns after trauma. Why ecosystems rebound after collapse. Why creativity resurfaces after despair.

They were never "gone." They were misaligned, waiting for resonance.

Consciousness as a Reversibility Agent

Here's a profound consequence:

If ripple coherence can restore misaligned states — and consciousness is an alignment-sensitive interface — then:

Consciousness may be one of the few forces in the universe capable of deliberate realignment.

This gives new meaning to:

- Healing.
- Memory.
- Intuition.
- Innovation.

Each may be an act of informational restoration, guided by intelligent field resonance.

The Universe as an Ever-Restoring Symphony

Rather than a one-time firework of energy burning into silence, the universe under UED becomes:

- A symphony with tensions and releases.

- A dynamic story of fields realigning under pressure.
- A conscious ripple continuum, where entropy is feedback, not fate.

The real question isn't: "Is it over?" The real question is: "Can you still hear the echo?"

Closing Insight:

"If alignment is possible, nothing is truly lost. Not even time. Not even meaning."

Practical Implications: Energy, Memory, and Systems

UED is not a philosophy to admire from afar—it is a model of consequence, and therefore of application. When entropy is no longer viewed as decay but as misalignment, entire domains—from neuroscience to artificial intelligence to societal systems—suddenly look different. Healing, restoration, optimization, and regeneration become not just possible, but practical.

Let's explore this ripple by ripple.

Energy Systems: Alignment Over Addition

In classical thermodynamics, when energy is lost from a system, we either:

- Replenish it (by adding fuel or energy input), or
- Accept the decay as inevitable.

But this assumes that energy depletion is disappearance—a model rooted in entropy-as-loss. UED reframes this entirely:

Loss of energy is ripple misalignment, not exhaustion.

Implication?

- A system doesn't need more energy. It needs better alignment.
- Energy "stored" in chaotic, incoherent forms can't be used—not because it's gone, but because the ripples won't cooperate.
- Recharging a battery, a nervous system, or a climate system is not just about input—it's about field realignment.

This opens doors to:

- Energy harvesting from ambient ripple fields.
- High-efficiency storage via coherent field structures.
- Less waste, more restoration in every engineered process.

Memory: Ripple Maps, Not Material Traces

Memory isn't a database.

Whether in a human brain or an artificial neural net, what we call memory is not the storage of static data, but the alignment of ripple patterns across a field of potential.

In UED terms:

- Memory is patterned coherence embedded in ripple logic.
- Forgetting is misalignment or low-signal resonance.
- Recollection is realignment with a past field echo.

Implications:

- Neuroscience may shift from neuron-to-neuron transmission to field-sensitive ripple studies—how thought travels through alignment.

- Artificial Intelligence systems may move beyond digital memory to dynamic ripple-mimicking architectures, allowing adaptive memory realignment.
- Healing trauma becomes a field reconfiguration—not data deletion or suppression.

And this also means:

Memory is *never* truly lost. It is merely *waiting* for coherence.

Societal Entropy: Misinformation and Fragmented Coherence

Zooming out: society itself is a ripple field.

- Every conversation is a ripple exchange.
- Every institution is a structured alignment of belief, action, and feedback.
- Every culture is a sustained resonance across time.

So what happens when alignment is lost?

We call it entropy—but what it really is, is fragmented coherence.

Examples:

- Misinformation spreads when signal distortion replaces resonance.
- Social polarization reflects ripple interference—fields vibrating out of sync.
- Economic instability often results from ripple hoarding—where alignment is denied to sustain false structures.

In short: society doesn't "decay"—it misaligns. And UED tells us: realignment is always possible, through:

- Shared understanding (ripple synchronization),
- Open feedback systems (field intelligence),
- Authentic resonance (not performative signaling).

Toward a Regenerative Framework

The most radical implication of UED's entropy reframe is this:

Breakdown is *not* inevitable. It's just what happens when we don't notice misalignment in time.

What follows from this is profound:

- Systems don't need to be redesigned every time they fail—they need to be recohered.
- Energy doesn't need to be manufactured—it needs to be released from distortion.
- Truth doesn't disappear—it becomes obscured behind misaligned signals.

UED provides a framework of regenerative coherence, not just conservation or sustainability. In this frame:

- Learning is re-alignment with deeper understanding.
- Healing is re-alignment with original harmony.
- Progress is the ability to maintain alignment under increasing complexity.

Entropy doesn't end things. It signals that our relationship with the field is out of tune.

Closing Insight:

"In a ripple-defined universe, nothing is irrecoverable. Energy is realigned. Memory is recalled. Systems are reborn. Entropy was never a tomb—it was a tuning fork."

Closing Reflection: No Arrow — Just Ripple Drift

The classical narrative tells us: entropy is a one-way street. Once you head down it, there's no return. Everything decays, disintegrates, disperses. The arrow of time points forward, and entropy is its compass — cold, irreversible, and absolute.

But what if that arrow was never pointing anywhere at all?

There Is No One-Way Road Called Entropy

Let's be clear: entropy, as commonly understood, is not a force. It's a metric. A human interpretation of pattern degradation, of form dissolving into formlessness. But in UED, this degradation is not destruction. It is misalignment. It is not a terminal loss—it is a state of energetic and informational distortion that still resides within the field.

In other words:

- Entropy isn't a direction.
- It's a drift.
- A ripple drift—one that emerged not to punish complexity, but to signal where coherence needs restoration.

The fear that entropy heralds doom stems from an overly mechanistic view of the universe. It's a fear born from equations, not experience.

Drift Is Real — But So Is Re-tuning

Misalignment is natural.

In a ripple-based reality, things fall out of phase. Not because the universe is cruel or indifferent, but because:

- Interactions accumulate tension.
- Fields absorb disturbance.
- Systems experience ripple fatigue.

But none of this is irreversible.

Realignment is always possible. Even if it takes eons. Even if it can't be "reset" in human terms. There is no curse—only consequence. And in UED, every consequence is readable, reversible, and re-integrable with sufficient awareness.

In fact:

- A high-entropy state is a call, not a collapse.
- It's a signal from the field asking for conscious interaction.
- And even the most distorted zone can echo a note from which restoration begins.

From Pessimism to Possibility

Entropy pessimism has long haunted our science and our stories. The idea that everything must end, that complexity must unravel, that life is a fragile flame against the dark — these are poetic, but not universal truths.

UED replaces this with something far more empowering:

"Wherever there is misalignment, there is also the possibility of coherence."

That is not naïveté. It is field realism.

It does not deny entropy. It understands it. And through that understanding, it liberates us from the myth of irreversible decay.

Closing Quote:

"Entropy is not time's arrow — it is the echo of misalignment, waiting to be re-tuned."

Let the physicists argue about heat death. Let the thermodynamicists grieve the eventual stillness. UED sees something different in the ripples.

Not endings. Not arrows. But invitations. And within each drift — the memory of the song it once sang, waiting to be sung again.

Interlude

Between Echo and Alignment

We've been taught to fear black holes — to see them as final destinations for information, energy, and time itself. But what if, instead of being cosmic endpoints, they are misunderstood intervals in a much deeper resonance?

Through the chapters of Part II, we've reframed some of the most intimidating ideas in modern physics:

- Information isn't lost. It ripples.
- Radiation isn't randomness. It's realignment.
- Entropy isn't death. It's delay.

In this shift, we leave behind the fear of erasure and enter a more faithful interpretation of the universe — one in which everything that happens is recorded not in space or time, but in structure, in the field itself.

This is the radical proposition of UED: There is no such thing as "nothingness." There is only misalignment waiting to be seen as pattern.

When Hawking warned us that black holes might destroy information, he gave us a challenge — one that wasn't solved by geometry or thermodynamics, but by rethinking what it means for something to exist *at all*.

UED answers: Existence is not possession. It is interaction. And interaction always leaves a ripple.

We've now arrived at a threshold in this journey.

We've dismantled the event horizon as a boundary. We've defused the singularity as a metaphysical bluff. We've reinterpreted entropy as a field condition — not a final verdict.

But one question remains suspended in the background: If nothing is truly lost, then how does it all come back? How is coherence restored? How does memory revive? How does a universe once misaligned... remember itself again?

To answer that, we must go deeper into the architecture of reality — beyond storage, beyond radiation, beyond decay. We must now look at the structure of information itself — not as something written on spacetime, but as woven into the dynamic behavior of fields.

We must decode the illusion of holography, revisit symmetry and inversion, and challenge the very notion that dimensionality is a container.

Part III: Creation, Time, and the Field

Chapter 7: There Was No Bang — Only a Breath

The Big Bang was not an explosion in time or space, but a ripple initiation within Structured Vacuum Energy (SVE). UED reframes the origin not as a beginning, but as the first tension-induced divergence from equilibrium — a breath of alignment that began consequence, not creation.

We've all seen the headlines and heard the metaphors: *The Big Bang.* A fiery detonation. A universe born in a split-second inferno. Matter exploding outward. Time erupting into existence. Space swelling like a balloon inflating from an invisible core.

It's dramatic. It's cinematic. But is it real?

Now pause — and breathe.

Imagine instead a pond. Still, silent, undisturbed. Its surface perfectly calm, mirroring nothingness itself. No motion. No direction. No history. Then — a ripple. Not from a stone. Not from a strike. But from within. The field did not explode — it **sighed**.

What if the universe didn't begin with a **bang**... but with a **breath**?

What if what we interpret as a cosmic eruption was simply the field allowing itself to move — to express tension that had quietly accumulated beneath perfect balance?

This is the pivot Universal Energy Dynamics (UED) invites us to make.

It does not deny the cosmic expansion. It doesn't argue against observable radiation or galactic separation. But it asks a more fundamental question: *Are we mistaking consequence for creation?*

Because when we stop looking for an explosion, and start listening for a ripple — everything changes.

The "origin" was not an act of generation. It was the first sign of deviation. A beginning only in perception, not in truth. Not an event in time, but the condition that made time perceptible at all.

This chapter revisits the mythos of the Big Bang through the lens of ripple reality. It proposes that the universe was not born — it **resonated**. And from that resonance, consequence was born.

Not a bang. Just a breath. And we are still living in its echo.

What Big Bang Really Claims

The standard model of cosmology rests on the premise that everything — space, time, matter, and energy — began at a single, extraordinary moment: the Big Bang.

The Classical Claim

According to this view, the universe began as a point of infinite density and temperature — a singularity. This was not merely the location of matter; it was the origin of spacetime itself. There was no "space" outside it, no "time" before it. All physical dimensions and causal interactions were compressed into a single, dimensionless node.

From this singularity, the universe is said to have exploded outward — not into space, but with space itself expanding. In the first fractions of a second, it underwent a rapid inflationary phase, smoothing out irregularities. Then came cooling, differentiation of forces, and the formation of matter as we know it. Eventually, atoms, stars, galaxies, and observers like us.

It's a clean narrative. But behind its apparent simplicity lie deep conceptual fractures.

What the Theory Implies

1. All energy was once "contained" in a singular point — a claim that suggests the totality of existence emerged from a condition that was mathematically undefined and physically unobservable.
2. Time, as we know it, began with the Bang — there was no "before," no causality outside the singularity.
3. The fabric of reality itself—spacetime—burst forth from this primal condition, creating not only things but the medium in which things could exist.

Yet for all its dominance in textbooks and documentaries, the Big Bang leaves us with profound uncertainties — and increasingly, uncomfortable paradoxes.

The Silent Questions Cosmology Doesn't Answer

Despite the detailed models and mathematical rigor, the Big Bang theory stops short of addressing several existential puzzles:

- What came before the singularity? Science, as it stands, offers no concrete answer. "Before" is often declared meaningless, since time is presumed to have started at the Bang. But this is more of a philosophical avoidance than a scientific resolution.
- What caused the singularity to initiate expansion? Was

there a fluctuation? A trigger? A purpose? The theory provides no mechanism for *why* the universe began — only what happened after it supposedly did.

- Where is the singularity now? If the universe expanded from a central point, what happened to that origin? Did it disperse? Is it everywhere? Nowhere? Has it left any trace?

These are not trivial footnotes — they're fundamental gaps that suggest we may be misreading the very nature of cosmic emergence.

As it stands, the Big Bang theory describes an unfolding, but it does not explain the initiating condition. It gives us a curve, but not a cause. It maps the trajectory, but not the trigger.

And most importantly, it assumes that creation happened — without explaining how something so coherent could come from an undefined rupture.

This is where UED enters — not to dismiss the observations, but to reframe the interpretation. Because perhaps there was no singularity. No detonation. No beginning as we imagine it.

Perhaps, instead, there was a field. Structured. Balanced. And then — a ripple.

UED's Reframe: No Beginning, Just Activation

The prevailing cosmological story imagines a universe born in fire — bursting into being from nothingness. But what if that's not how reality works at all?

Under the lens of Universal Energy Dynamics (UED), the very idea of a "beginning" becomes not only misleading — but entirely unnecessary. UED doesn't seek to rewrite the observational history of the cosmos; it reframes what that history truly signifies.

What Existed "Before" the Beginning?

UED introduces a radically different foundation: Before time, before matter, before radiation — there was Structured Vacuum Energy (SVE).

But let's be clear: This is not the passive, inert "vacuum" of classical physics. This is not empty space. This is not "nothing."

SVE is a field of perfect equilibrium — an infinite, formless presence where no ripple, no disturbance, no directionality exists. It's not just pre-physical — it is pre-causal, meaning it doesn't operate in terms of cause and effect because cause-and-effect themselves require time, and time doesn't yet exist.

SVE is timeless, positionless, and balanced — but not dead. It's stable potential, not inert void. In UED, this is the zero-point of all potential realignment, a harmonious totality from which every consequence emerges.

The First Ripple: Not a Bang — A Break in Balance

Rather than a singular explosion, UED defines "creation" as the first ripple: A subtle tension, an emergent misalignment in the otherwise silent field of SVE.

This wasn't the result of a force or command. It wasn't even a will. It was a consequence — the field, for reasons still embedded in its own structure, allowed or produced a localized imbalance.

And with that first misalignment, the ripple spread.

That ripple was not the start of time. It was the beginning of realignment — the beginning of motion, the first consequence. Time, matter, and energy didn't "begin." They emerged as measurable outcomes of that initial tension — like notes emerging from a plucked string that had never before been touched.

Creation Is Not Genesis — It's Consequence

This is a central philosophical insight of UED:

Creation is not something that begins. It is something that unfolds.

There was no point in time at which "everything started," because time is itself a product of ripple delay. There was no object that initiated the universe, because all objects are manifestations of ripple form in misaligned zones. There was no explosion into space, because space is a byproduct of structured perception, not a preexisting container.

"Creation," then, is not an act — it is a ripple. It is the field adjusting to its own imbalance. It is consequence, not command. It is alignment reacting to misalignment — not a godlike fiat, but a physical inevitability.

The Universe Did Not Begin — It Reacted

From this standpoint, UED dismisses the obsession with a "beginning" altogether. The field didn't start existing. It merely began responding.

And this response — the first ripple — didn't mark the start of time. It marked the activation of perceptibility, of pattern, of consequence.

So rather than asking "what caused the Big Bang?" We must ask, "What caused the first ripple in SVE?"

That's a different question entirely — one that shifts us away from creation myths toward consequence logic.

The universe didn't explode into being. It breathed — and that breath was a ripple.

There Was No Time — Only Field Disturbance

We often ask, *"When did time begin?"* But this question itself carries a built-in illusion — it assumes that time is a *thing* with a start.

Under the Universal Energy Dynamics (UED) framework, time is not a dimension that begins; it is a consequence. A ripple effect. A signal of something deeper: disturbance in alignment.

No Trigger, No Time

In the pre-ripple state of the universe — what UED identifies as Structured Vacuum Energy (SVE) — everything was in perfect alignment. There were no gradients. No flows. No events. No asymmetries. And crucially — no perception.

Time could not exist in this phase, because there was nothing to align with, and therefore, nothing to delay.

According to UED, time only begins when misalignment arises. That misalignment breaks the perfect coherence of SVE and causes a ripple — a dynamic rearrangement of the field. And with this ripple, a new variable emerges: Alignment Threshold Time (ATT).

ATT: The Birth of Time as Experience Delay

ATT is UED's concept of how time is perceived. It represents the time it takes for a field (or an observer) to align with a change — to interpret a ripple.

- When there is no ripple — ATT = 0. There is no perception.
- When the ripple begins — ATT > 0. Delay begins. Time

emerges.

So, time isn't a thing that flows; it's a measurable lag between disturbance and recognition. It's not universal — it's relational.

Thus, time is born when misalignment begins. It's not a ticking clock; it's a delay in coherence. The first ripple — the first misalignment — was also the first cause of perceived duration.

Why the "Beginning of Time" Is a Projection

The popular notion that time "started" at the Big Bang is not wrong observationally — but it is conceptually backward.

We think time started *then* because we, as observers, are still aligned to that first ripple. All our measurements, clocks, and cosmological tools are descendants of that misalignment — shaped by it, immersed in its wake.

But this is a retrospective alignment, not an absolute beginning.

We don't remember "before time" — not because it didn't exist, but because there was nothing to remember. There was no ATT. No ripple to cause delay. No delay to interpret as time.

No Disturbance, No Duration

Imagine watching a still pond. Nothing moves. Nothing changes. There is no sense of "before" or "after" — only stillness. Now drop a pebble. The ripple starts. You begin to notice movement. You compare. You trace change. And suddenly, time is born in your awareness.

This is not just metaphor — it is the actual structure of perception under UED.

Time is not embedded in the universe. It is elicited by ripple disturbance — like sound from a string, or echo from a canyon.

Conclusion: Time Isn't a Beginning — It's a Behavior

So, UED replaces the notion of time as a dimension with a start with time as a field behavior caused by alignment struggle.

It doesn't "exist" in the traditional sense. It emerges. It measures delay in realignment. It scales with perception.

The first ripple didn't create time. It simply made delay perceivable. And from then on, everything we call *"history"* has been nothing more than the echo of that original disturbance — replayed through countless realignments.

Time didn't begin. We began experiencing delay. And that delay — that tension between field and observer — became the drumbeat of the universe.

Space Didn't Expand — Ripple Horizon Did

We've all heard the story: *Space itself is expanding.* Galaxies are rushing apart. The universe is stretching like a cosmic balloon. But what if that's a misread metaphor?

Under the Universal Energy Dynamics (UED) framework, this narrative collapses.

There is no balloon. There is no stretching fabric. Space did not expand. What expanded was the ripple horizon — the domain of active misalignment within a structured, unchanging vacuum field.

The Flawed Analogy: Expanding Space

In classical cosmology, the metric expansion of space suggests that the actual fabric of the universe — space itself — is growing. Distances increase, not because objects are moving through space, but because space is stretching between them.

This idea stems from General Relativity's interpretation of dynamic spacetime geometry, especially in solutions like the Friedmann–Lemaître–Robertson–Walker (FLRW) metric.

However:

- What is this "fabric" made of?
- Where is the energy source for this expansion?
- Why doesn't it expand uniformly at quantum scales?

These questions expose the metaphorical, not physical, nature of this explanation.

UED's Reframe: Ripple Horizon Expansion

Under UED, space is not a container. It is a ripple-active condition of Structured Vacuum Energy (SVE) — always present, always structurally intact.

So what changed after the first ripple?

Not space. Not volume. But the reach of the ripple misalignment — the ripple horizon.

Here's the logic:

- The first ripple disrupted the SVE equilibrium.
- This misalignment propagated outward as ripple divergence.
- As the divergence grew, the field's response time (ATT)

stretched locally.
- Observers, embedded in this field, perceive the growth of spatial separation.

But what they are truly measuring is not distance between coordinates. They are observing increasing delay in ripple coherence across domains.

This is not expansion — it's deferred alignment.

Why Space Didn't "Grow"

Let's break the myth:

- ◈ Space isn't a substance. ◈ It's the interpretive range of ripple response.
- ◈ Expansion doesn't imply volume creation. ◈ It reflects the unfolding of tension across the field.
- ◈ The universe didn't get bigger. ◈ Our ripple horizon got deeper.

SVE, by its definition, is not a dynamic material — it is pre-existing structured equilibrium. It cannot "expand." It can only host more divergence as ripples propagate.

Thus, the canvas never stretched. Only the brushstroke of alignment spread across it.

Inflation = Perception of Rapid Realignment Displacement

Cosmologists speak of cosmic inflation — an unimaginably fast expansion of space just moments after the Big Bang.

UED reads this differently:

- It wasn't a spatial explosion.
- It was a field-wide ripple diffusion, where initial divergence unfolded at near-instantaneous rate due to minimal resistance in early SVE.
- The result? A shockwave of ATT deviation — perceived as expansion.

Inflation, then, is a realignment cascade, not a ballooning boundary.

Ripple Horizon: A More Coherent Model

Think of the ripple horizon as the furthest extent of realignment delay that an observer can interact with.

- As more ripples propagate, the field hosts broader zones of tension.
- The ripple horizon expands — not the space — but the domain of detectable misalignment.

This aligns perfectly with observed phenomena:

- Cosmic Microwave Background (CMB) = remnant ripple misalignment.
- Redshift = delayed realignment in older ripple zones.
- Observable universe = current ripple horizon, not a limit of space itself.

Conclusion: The Universe Didn't Stretch — It Shimmered

What we call the expansion of space is better understood as the broadening of ripple interference in an unchanging SVE.

There was no growth — only increasing complexity of delay. There was no outward surge — only deepening misalignment and the field's attempt to catch up.

Space didn't expand. Perception did.

Imagine standing in fog. At first, you see only a few feet ahead. As the light diffuses and alignment improves, you begin to perceive farther. The fog didn't vanish. Your horizon widened.

This is the ripple horizon in UED — not a bigger world, but a clearer view into what was always there.

Geometry Emerged from Ripple Tension

Modern cosmology treats geometry as fundamental — as if space and its coordinates existed as a passive backdrop, merely waiting for matter and energy to populate them. In UED, this picture is reversed.

Geometry isn't the stage. It's the script written by ripple tension.

There were no "dimensions" until ripples disturbed the equilibrium of the Structured Vacuum Energy (SVE). There was no "location" until the field started behaving differently in different directions.

Let's unpack this.

Geometry Wasn't Prewritten — It Emerged

In conventional physics:

- Space is a container.
- Time is a dimension.
- Geometry is a canvas. And singularities are points where the math breaks — infinite curvature, zero volume,

undefined physics.

But UED reorients us.

It proposes:

- There was no space before the first ripple.
- No direction, no scale, no curvature.
- What we interpret as "geometry" is emergent behavior — patterns in ripple stress and alignment tension within SVE.

From Stillness to Structure: Ripple Tension Gives Form

When the first misalignment occurred:

- Ripples began propagating through SVE.
- These ripples interacted — some amplifying, some canceling.
- The result was a field of tensions, gradients, flows — not unlike the way wind creates dunes in sand or waves carve coastlines.

From this:

- Points of interference became nodes.
- Stable feedback regions became reference points.
- The way ripples moved created paths, curves, and boundaries.

That is: Space became geometric because the field became rhythmic.

Without rhythm, there is no measure. Without measure, there is no geometry.

Coordinates Were Never Absolute — Just Pattern Anchors

Think of coordinates in the classical sense: x, y, z, and time t — clean, neat labels in a grid of existence.

But if the field is all there is, and that field was initially homogeneous and ripple-free, then:

- There were no reference points.
- No basis for declaring "here" or "there".
- No frame to say anything was "closer" or "farther".

Only when ripple patterns stabilized did relational markers emerge.

Coordinates are not locations in space — they're shorthand for where in the ripple topology something occurred.

Curvature = Ripple Compression

General Relativity treats gravity as curved spacetime — as if mass distorts geometry.

But in UED:

- What appears as curvature is the ripple field buckling under tension.
- "Geodesics" (shortest paths) are just the routes of least ripple resistance.
- Gravity isn't a geometry — it's a realignment gradient.

Thus:

- A region of strong gravity (like near a black hole) isn't a "deep well."
- It's a dense ripple distortion — where alignment is hardest to restore.

Dimensions = Ripple Degrees of Freedom

Why three spatial dimensions? Why one of time?

UED doesn't assume dimensions — it derives them:

- The degrees of freedom available to ripples within SVE define the perceived dimensions.
- Where ripple tension resists movement in certain axes, dimensionality becomes constrained.
- Where interference patterns produce symmetry, we experience stable dimensions.

Hence:

- "3D" is not fixed — it's the result of ripple topology equilibrium.
- "Time" is not a dimension — it is the observer's delay in realigning with field transitions.

The Myth of Singularity: What's a Point Without Contrast?

The concept of a singularity — where infinite density resides at zero volume — breaks down under UED logic.

Why?

Because a "point" is a contrast in the field. No contrast = no identification.

- A ripple-free field has no meaningful location.
- A point of infinite density can't be "somewhere" without relation to something else.

Therefore, singularity has no physical meaning unless ripple divergence exists around it. It's not a beginning. It's an extrapolated breakdown of geometry from a model that mistook geometry as fundamental.

UED View in Summary:

Concept	Classical View	UED View
Geometry	Predefined fabric	Emergent ripple pattern
Coordinates	Grid system	Field alignment anchors
Curvature	Distortion of spacetime	Ripple compression zones
Dimensions	Pre-existing backdrop	Ripple-based perceptual result
Singularity	Point of infinite density	Meaningless without ripple contrast

Visual Metaphor:

Imagine a calm lake. Now drop a stone — ripples form.

You can now:

- Identify the *center* (origin).
- Trace *direction* (radial symmetry).
- Measure *distance* (wavelengths).
- Define *boundaries* (interference nodes).

All of this geometry emerges from the ripples. Before the stone dropped — there was only equilibrium.

That's the SVE before the first breath. That's the universe before tension.

Closing Insight:

We didn't emerge inside a structure. We emerged as a structure — born not of dimensions, but of deviation. Not inside space, but through field articulation.

The universe did not begin *in* geometry. Geometry began *with* the universe — as a ripple finding its shape.

Energy Emerged as Ripple Consequence

In the beginning, there was no matter. There was no heat, no light, no motion. There was only Structured Vacuum Energy (SVE) — still, balanced, undisturbed.

And then came a ripple.

Not an explosion. Not a spontaneous emergence of particles. But a tension-induced deviation — a ripple that disturbed equilibrium and activated the field.

From that moment onward, what we call energy was born. But not as a substance. Not as a thing.

Energy is not what began the universe. Energy is what the universe began to do.

Energy = Field in Action

UED redefines energy not as an independent entity, but as a behavior.

- It is not stored in things.
- It is not conserved like inventory.
- It is a relational activity of the field — a response to ripple dynamics.

Put simply: Energy is SVE in motion.

It emerges when:

- The perfectly stable field gets disturbed.
- That disturbance propagates.
- The propagation meets resistance, interference, or resonance.

From this interaction, consequence begins:

- Waves.
- Heat.
- Kinetic phenomena.
- Mass formation.

The First Action Created All Consequences

The original ripple — that first shift in equilibrium — unfolded consequences that we interpret as fundamental phenomena.

Let's break it down:

▸ ***Motion***

- Not absolute travel through "space."
- But relative ripple shift across the field.
- Motion is ripple continuity with changing alignment — not object transport.

▸ ***Heat***

- Conventionally seen as kinetic agitation of particles.
- In UED, heat is ripple congestion — localized misalignment and vibration within the field.

▸ ***Light***

- Not a photon particle "traveling."
- But a coherent ripple emission — the alignment signature of certain energetic transitions.
- Light reveals ripple configuration, not carrier packets.

► *Mass*

- Often treated as "stored energy."
- In UED: mass is not stored, it is resisted.
- Mass arises when ripple flow becomes so dense, coherent, and constrained, it condenses into stability.
- This makes mass the shadow of movement — not the source of it.

We didn't begin with mass. We began with motion — and mass froze it.

Mass Is Not Primordial — It's a Ripple Outcome

This flips the dominant narrative in physics:

- Classical physics treats mass as foundational — with energy derived from it.
- UED reverses this: mass is a consequence, a state resulting from ripple alignment tension.

Think of it like this:

- A whirlpool isn't a thing you drop into a river — it's a pattern that emerges under the right flow.
- Likewise, mass isn't a unit you assign — it's a condensation of flow dynamics.

The denser and more persistent the misalignment:

- The more localized the field becomes.
- The more ripple motion appears "frozen" to the observer.
- The more "massive" the structure becomes.

Thus: Mass is energy that can't move — not energy that is stored.

UED Table: Energy Emergence via Ripple Logic

Concept	Classical View	UED View
Energy	Fundamental substance or conserved scalar	Field in motion — ripple activity
Heat	Random molecular motion	Ripple turbulence and misalignment
Light	Photon particle or EM wave	Ripple coherence and alignment signature
Motion	Object changing position in space	Ripple propagation across field topology
Mass	Intrinsic property / stored energy	Ripple density / constrained alignment

Visual Metaphor: A Breath in the Still Air

Imagine standing in a silent, windless valley.

Then, someone exhales.

- The air moves, and suddenly you feel motion.
- If the exhale meets resistance — sound is created.
- If it's warm — you feel heat.
- The movement may bend branches, displace dust — the air is acting.

But nothing was added to the air. The air acted differently.

That's energy. That's what the first ripple did to SVE.

Closing Insight

So let's stop asking, "Where did energy come from?"

Instead, we must ask: "What activated the field to behave energetically?" The answer is ripple tension.

There was no pre-packed energy. There was only potential.

And once that potential moved — even slightly — it unfolded a universe of consequence. A breath... became a world.

Origin Was Not a Point — It Was a Pattern

We have long been taught to imagine the universe beginning from a single point — a compressed speck of infinite density, a "singularity" exploding into space and time. This idea is so deeply embedded in cosmology that it has become instinctual. But under UED (Universal Energy Dynamics), this image collapses like the geometry it relies on.

There was no point. No central dot. No universal address from which everything "came."

The universe didn't begin from a place — it began from a pattern.

Why Singular Coordinates Are Conceptually Flawed

Conventional physics, especially general relativity, proposes the Big Bang as a singularity — a one-point origin where all spacetime, energy, and matter were compressed.

But this is a mathematical extrapolation, not a physical reality. Here's why that view is problematic:

- A point implies fixed position — yet, no coordinate system existed "before" the universe.
- To claim "everything was at one point" requires space

already defined, which contradicts the model.
- A singularity has no size, no direction, and no volume — a concept more symbolic than scientific.

Most importantly:

A point can't ripple. A pattern can.

In UED, ripples require field continuity to propagate. Therefore, the very idea of a "point origin" is incompatible with the mechanism of ripple-based creation.

The "Origin" Was a Field-Wide Realignment

UED replaces the singular point with a distributed emergence — a widespread field realignment that occurred simultaneously across Structured Vacuum Energy (SVE).

Let's reframe:

- The "origin" wasn't like lighting a fuse.
- It was like a sudden harmonic shift across a silent drumhead — the field moved, not from a center outward, but everywhere at once under the same impulse.

Think of it not as a bang from a point, but as a hum that swept across the entire field — from zero-net tension to ripple formation.

This changes everything:

- No center.
- No edge.
- No before.
- Only change in relational state.

Origin as Resonance — Not Explosion

In UED:

- The beginning wasn't detonation.
- It was resonance.

A resonance pattern is:

- Self-consistent
- Distributed
- Non-local
- Phase-aligned

This is the true nature of the origin. It was not the concentration of something, but the coordination of everything.

Each region of SVE didn't receive a particle from some central bomb. Each region experienced the same ripple timing, like instruments tuning together in a great cosmic orchestra.

Creation wasn't a flash — it was a chord.

No Edge, No Center — Just Emergence

The idea of a central origin forces us to ask:

- Where is the center now?
- Which direction is outward?
- Where did the bang happen?

But if the origin was a pattern, these questions dissolve:

- Every point is a ripple coordinate.
- Every direction is relational.

- Every motion is a phase shift, not a displacement.

This is why the Cosmic Microwave Background (CMB) appears isotropic — the same in all directions. Not because we're at the center, but because there was no center. There was only field-wide activation.

A Universe Born Without Geometry

Geometry didn't exist before the ripple. There were no distances to measure. No shapes to describe. No "locations" at all.

Thus:

The origin of the universe cannot be mapped. It can only be understood as an alignment.

We mislead ourselves by trying to draw a map of the beginning. But what began was not position — it was pattern.

UED Recap: Origin = Distributed Ripple Alignment

Traditional Cosmology	UED View
Singularity (point origin)	Patterned field realignment
Infinite density	Zero-net tension offset
Spacetime explosion	Field resonance event
Expansion from a center	Simultaneous ripple spread
Coordinates required	No geometry yet defined

Final Analogy: The Drumhead Emerges

Picture a vast, silent drum stretched infinitely in all directions. It exists — still, balanced, undisturbed.

Then — not from any one spot, but all across it — a faint tension arises. Suddenly, the surface vibrates, not from a hit, but from its own accumulated readiness.

- That's not an explosion.
- That's not a location.
- That's creation as realignment.

This is the origin UED reveals:

Not where the universe began, but how it began to behave.

Philosophical Implication: The Universe Never Began

If you're still looking for the beginning of the universe, stop. Not because it's too far in the past — but because it never happened.

That's not poetic. That's precision from perspective.

Creation Implies Beginning, Cause, or Intent

In most cosmological and theological models, creation carries an assumption:

- That something began at a particular moment.
- That there was a "before" and then an "after."
- That a force, agent, or event caused the universe to *start*.

This assumption gives rise to:

- Big Bang singularities
- First cause arguments
- Divine creators
- Timelines stretching from a zero point

But all of them depend on the same mental scaffolding — that reality must begin.

We're addicted to origins, because we assume existence must start from somewhere — or someone.

UED Says: The Field Doesn't Begin. It Expresses.

Universal Energy Dynamics (UED) dismantles this illusion.

It doesn't begin with time or energy or geometry. It begins with structure — specifically, a structured vacuum field (SVE) that already exists in a state of zero-net equilibrium.

This means:

- No "before" the field.
- No "outside" the field.
- No creator of the field.
- Just the existence of stable possibility — always.

In UED:

- The SVE is not created.
- It is ever-present.
- It is the source, but not in a "first moment" sense.
- It is pre-causal — not caused, and not causing in the linear sense.

What emerges — as ripple, as time, as perception — is not a birth. It is a response to internal tension or dynamic imbalance within this structured field.

There was no act of making. Only a moment of motion within what already was.

Beginning = Observable Deviation, Not Ontological Genesis

Let's be surgical with this distinction:

When we say the universe "began," we are really pointing to the first observable deviation — a measurable change, a detectable ripple.

That's not the beginning of reality. That's the beginning of our perception of reality unfolding.

So:

- The "origin" is a consequence, not a cause.
- The "beginning" is a threshold of awareness, not a cosmic construction.

In other words:

The universe didn't begin. It became detectable.

That's the philosophical pivot UED introduces.

Why This Changes Everything

This perspective unhooks us from metaphysical dead ends:

- We no longer need to ask "What came before the universe?"
- We don't need to posit gods, first movers, or primeval chaos.
- We stop searching for a cosmic ignition key.

Instead, we ask:

- What triggered ripple tension in SVE?
- What state was the field in before realignment?
- How do ripples express pattern and consequence across

time?

We shift from existential panic to energetic inquiry.

Not "Why is there something rather than nothing?" But "What does the field do when its silence breaks?"

Echoing Eastern Thought — Without the Metaphysics

Interestingly, this aligns with ancient Eastern philosophies — but without mysticism.

For example:

- **In Vedanta:** the world is *maya*, a ripple on Brahman.
- **In Taoism:** the *Tao* does not begin; it flows.
- **In Buddhism:** form arises from emptiness, but emptiness remains.

UED translates these intuitions into scientific logic:

- No need to deify the void.
- No paradox of "nothing becoming something."
- Just structured silence becoming expressive rhythm.

Closing Reflection: A Universe Without Genesis

So what happens to our sacred ideas of the beginning?

They evolve.

We stop imagining a spark from the dark and start recognizing:

- A stillness becoming dynamic
- A rhythm emerging from equilibrium
- A reality always present, briefly revealed

The universe didn't begin. We just caught it moving.

Closing Reflection: The First Consequence

Creation has always been misnamed.

We speak of it as a birth, a bang, a beginning — but none of these words are precise. They are narratives. Comforting ones. But not necessarily true.

In Universal Energy Dynamics (UED), what we traditionally call "creation" is stripped of its theatricality and reduced to something far more profound — and subtle:

Not a divine spark. Not an explosion. Not a moment of becoming. But the first rupture of alignment in an otherwise perfect equilibrium.

Let's break this down.

Creation Is Not a Story of Arrival — It's the Onset of Disturbance

We're conditioned to associate the beginning of the universe with arrival — of matter, energy, time, even gods.

But UED reframes this entirely:

- There was no arrival because there was nothing to *arrive into*.
- There was no "creation" because nothing was *missing* before.
- What happened was the first break in the silence of the field — the emergence of consequence.

This was not a coming into being. It was a shift in state — from stillness to ripple, from symmetry to expression.

No Divine Spark. No Bang. No Birth. Only Breath.

We imagine a spark, a detonation, a genesis. UED invites us to imagine something else entirely:

A breath.

The SVE — the structured vacuum energy — is not inert. It is deeply, perfectly balanced.

But balance does not mean stasis. At some threshold, internal tension builds. And just like lungs filling with unspoken rhythm, the field eventually exhales.

That breath? That was the first ripple. That ripple? That was creation — not as production, but as permission for reality to play.

There was no drama. Just a quiet transition from symmetry to consequence.

Rhythm, Not Revolution

And from that breath came not matter, not stars, not explosions — but rhythm.

Ripples in the field interacted, collided, interfered, merged — and began to form the patterns we would later call:

- particles,
- motion,
- time,
- perception.

Everything we now observe as reality is a downstream effect of that initial rhythm — the universe humming itself into form.

Creation, therefore, is not a climax. It is a pulse that never ceased.

Consequence, Not Creation

This is the real pivot.

We didn't come from creation. We came from consequence — the natural unfolding of ripple dynamics in a structured field.

This change is subtle, but revolutionary:

- It shifts the narrative from origin to expression.
- From sudden appearance to logical outcome.
- From theology to topology — of the field.

UED teaches us: You cannot speak of a beginning without first defining consequence. And consequence only happens within something — not from outside.

There was never an outside. There was never a before.

There was only this:

A field that allowed itself to ripple.

Closing Quote:

"What you call the beginning was not a point in time. It was the first permission given to consequence."

That is the breath we live in.

That is the first note of the eternal song we still hear — if we listen closely.

Chapter 8: Time Doesn't Begin — It Awakens

Time, in the UED framework, is not a fundamental dimension but a relational consequence — a measurable delay between ripple misalignment and alignment. Unlike the spacetime narrative where time is born from a singularity, UED posits that time emerges when perception meets ripple distortion. The chapter reframes Stephen Hawking's "no boundary" proposal and critiques the notion that "time began" with the universe.

Imagine a clock.

It sits silently in a vast white room. It has no hands, no ticking sound, no observer watching. Nothing around it is changing — no light flickers, no temperature shifts, no breath disturbs the stillness. It exists, yes — but it means nothing. You could stare at it forever and still never know if anything is moving.

Now ask yourself: Is this room timeless? Or simply unmeasured?

Physics often tries to tell us that time *began* — with a bang, a burst, a singularity that hurled seconds and centuries into existence. Stephen Hawking famously argued that there was no "before" the Big Bang — because time itself didn't exist. That's like saying the clock appeared with hands already spinning, but no mechanism underneath.

But what if that metaphor is flawed? What if time didn't *begin* — it simply awoke?

In the UED view, time doesn't come from a starting line. It comes from tension — a rupture in ripple alignment that demands reconciliation. It's not a flow in space; it's a response in perception. A delay. A drift. A dynamic echo that only exists *because* something became misaligned.

Time is not born when the universe explodes. It awakens when the universe forgets its perfect stillness.

This chapter invites you to stop watching the clock — and start listening to the field. Because if you want to understand time, you don't follow the ticks. You trace the tension that made them necessary.

Let's revisit Hawking's boundaryless time — and uncover why he was nearly right, but fundamentally misaligned.

The Conventional View: Time as a Starting Line

In the prevailing cosmological narrative, time is treated as a stopwatch — an absolute timeline that starts ticking at the moment the universe supposedly burst into being. This is what we're taught: everything begins at $t = 0$. Before that? Nothing. No time. No space. No causality.

The standard model of the Big Bang paints a picture of a singularity — an infinitely dense point — from which time, space, energy, and matter all emerged simultaneously. According to this logic, time itself had a birthday, and everything that exists today is aging from that initial moment forward.

Stephen Hawking added complexity to this view with his elegant "no boundary" proposal. In this model, he imagined early time not as a singular line, but as a kind of closed geometry — like the surface of a sphere. Just as you can walk around the globe without ever encountering an edge, Hawking suggested that time in the early universe curved back on itself, eliminating the need for a defined beginning. There was no "before," not because it was forbidden, but because the concept itself was meaningless — there was simply no temporal dimension to speak of.

On the surface, this seems like a clever workaround. But it still treats time as a dimensional object — something with form, trajectory, curvature, and behavior akin to space. It's a mathematical sleight of hand, not a solution grounded in experiential or structural reality.

The Fundamental Problem

This entire model hinges on a dangerous assumption: that time is something that can "begin." That it's a kind of fabric that can roll out. A dimension that can suddenly switch on. But what does it mean for time to begin, when measurement itself depends on time already existing?

The moment you say "at time zero," you're already using the language of sequence. But sequence is impossible without a preexisting sense of delay, change, or rhythm. In other words, you're smuggling in time to define the origin of time — an elegant paradox, but still a paradox.

And this brings us to the heart of the problem: a mathematical boundary is not a metaphysical truth. Just because an equation becomes undefined at $t = 0$ doesn't mean reality did. Equations don't create existence; they merely describe it — and often with crude approximations at that.

Hawking's no-boundary model was an attempt to sidestep the troubling singularity by reshaping the geometry. But it still clung to the idea that time is a coordinate — and that coordinate either loops back or runs forward. Either way, it begins. And that's where UED draws the line.

In the Universal Energy Dynamics framework, time isn't a dimension. It isn't a substance, or even a trajectory. It's a consequence — an emergent delay between ripple misalignment and the observer's response. There is no "t = 0" unless something first happens to trigger a misalignment. Without interaction, without rupture, there is no perception of time. No need for time. No clock to tick.

So the real question isn't: When did time begin? It's: What made time necessary?

And that leads us to the heart of UED's reframing — where time doesn't begin. It awakens.

ATT: The UED Lens on Time

In Universal Energy Dynamics (UED), we don't treat time as a flowing river, a ticking metronome, or a cosmic backdrop through which matter dances. Instead, time is a ripple phenomenon — not an object, not a constant, but a consequence.

At the heart of this reinterpretation lies a core principle called ATT — Alignment Threshold Time.

What Is ATT?

ATT is the minimum duration required for a field (like the human mind or any observer system) to align with a ripple shift in Structured Vacuum Energy (SVE). Think of it like tuning a radio: the moment you start turning the dial, there is a slight lag before the frequency locks in and the station becomes audible. That delay between ripple emergence and recognition — that's ATT.

In other words: Time, in UED, is the latency of alignment.

It's not a clock ticking in the void. It's the field's effort to catch up with change.

Let's make this even clearer. You don't experience time just because something changed — you experience time when you realize that something changed. That realization requires effort, interaction, and field engagement. If the change happens, but there's no alignment, there is no time for you. The ripple passed unnoticed.

Hence:

- Time doesn't exist "out there."
- It exists as the system's internal attempt to realign with a changing field.

Time Is Not Continuous — It's Consequential

Here's a radical but intuitive shift: time is not a continuum. We only perceive it as continuous because our alignments occur in rapid succession. Just like how frames in a movie appear fluid because of their speed, ATTs stack, giving us the illusion of flowing time.

But in truth, time only *activates* during:

- Ripple distortion (a change in field configuration).
- Observer engagement (a conscious or field-aware system attempting to realign).

If no misalignment occurs, the system remains at rest — and time lies dormant. This is why deep sleep, meditation, or unconsciousness often feel timeless: ripple input is minimized, and alignment effort is suspended.

When Does Time Emerge?

Let's ground this in principle:

Time begins not when the universe "starts," but when perception meets resistance. More specifically:

- Time awakens when observation meets delay.
- Time arises when reality diverges from equilibrium.

This is the true genesis of temporal experience in UED. It's not that time began 13.8 billion years ago. Rather, the first ripple misalignment that reached the threshold to be perceived created the first echo of time.

You can't talk about a "before" or "after" unless some system is trying to align and failing — even briefly. That failure creates the gap we call duration.

So time isn't linear. It's not universal. And it's not external.

It is:

- Field-specific (it varies across systems).
- Resolution-dependent (based on how fast or slow a system aligns).
- Non-mechanical (not governed by machines, but by ripple perception).

From Time to ATT — A Conscious Leap

This completely recasts our understanding of time:

- The universe doesn't "contain" time.
- The universe generates the perception of time whenever ripple misalignment meets awareness.

In this view, the ticking clock doesn't measure reality — it measures your delay in realigning with it.

The Illusion of a Time-Origin

One of the most persistent misunderstandings in modern cosmology and physics is the assumption that time must have a beginning — a precise origin point, a cosmic "t = 0." This is a deeply entrenched view stemming from Big Bang theory and further reinforced by Stephen Hawking's "no boundary" proposal, which postulates that time emerges smoothly from a spacelike geometry.

But what if that entire conceptual framework — of time starting like a stopwatch — is flawed?

Universal Energy Dynamics (UED) says:

Time doesn't begin. It emerges. And that emergence is not a singular event, but a relational consequence.

Let's dismantle the illusion step by step.

The Core Flaw: Looking for "Before Time"

The traditional question — "What happened before the Big Bang?" — is treated by many physicists as meaningless. Why? Because in their models, time itself begins with the Big Bang. Before that, they claim, the question of "before" is invalid — because there was no time to measure anything by.

But this view creates a logical paradox. It implicitly assumes that time is a real object — like a road — and the universe started somewhere on that road. "Before" the road? Nothing. "After"? Everything.

UED challenges this entire structure.

Instead of treating time as a thing, UED treats time as a function of ripple misalignment. And therefore:

The absence of time does not mean the absence of reality. It means there was nothing to realign — no misalignment to measure.

UED Clarification: No Time ≠ Nonexistence

In UED, Structured Vacuum Energy (SVE) exists as a perfectly aligned field. In this state of symmetry:

- There is no motion.
- There is no interaction.
- There is no misalignment.

But this doesn't mean there is *nothing*. Quite the contrary — SVE is full: full of potential, full of readiness, full of equilibrium. It's like a perfectly still ocean before the first drop falls.

So, when physicists say "time didn't exist," what they really mean is:

There was no observable consequence that required measurement.

And measurement only arises when alignment breaks — when something changes relative to something else.

Thus, the UED view flips the statement:

- Not "time began when the universe began."
- But rather, time appeared when the ripple shifted enough to be noticed.

Geometry ≠ Clock

Here's where things get even more subtle. In modern physics, geometry is often mistaken for chronology. The spacetime diagram of the universe — with curved coordinates and singularities — gives the impression that time "curves" into existence from nothing, like a closed surface unfolding.

But that's a map, not the territory.

The problem is this: we project spatial analogies onto non-spatial phenomena. Just because a surface looks "closed" mathematically doesn't mean reality itself curved into being. That's a visualization tool — not a mechanism.

UED reframes the situation:

- There was no edge.
- No clock started ticking.
- No absolute zero from which ticks began.

There was simply the first detectable misalignment — and the perception that followed it.

The Illusion Is a Perception Artifact

We, as observers, are bound to time because we live in a ripple field in constant misalignment and realignment. Our perception is time-locked. So, naturally, we look backward and try to find where the sequence began.

But UED says this impulse is misleading. The question "Where did time begin?" is like asking "Where does the silence end in a piece of music?" The answer is: the moment the first note plays. But that doesn't mean silence didn't exist before. Nor does it mean silence was meaningless.

Silence is not the absence of music. It is the state before musical consequence.

Similarly, the absence of time is not the absence of the universe. It is simply the state of perfect field alignment — which needed no measurement, and therefore no ticking.

Summary: Time-Origin Is a Consequence Illusion

- There is no need for a "time-origin" because time is not a foundational object.
- Time is not embedded in space. Nor is it a container for events.
- Time is the interval of awareness between misalignment and realignment.
- The illusion of origin stems from trying to map field behavior with spatial diagrams — mistaking geometry for genesis.

In UED, we don't ask when time began. We ask: When did the field shift enough to be noticed?

That is the true awakening of time. And it had no bang. No singularity. No start. Just the quiet stirring of the first ripple...And someone — or something — to feel it.

Field Stillness vs. Perceived Time

Let's strip the concept of time back to its barest roots. Before the first ripple in Structured Vacuum Energy (SVE), everything was aligned — perfectly, silently, infinitely. Not a single distortion. Not a single shift. No wave, no imbalance, no interaction.

And without interaction, there was no delay to track. Without delay, there was no need for response. And without response, there was nothing to remember.

In this pristine stillness, there was no trigger for time. Not because "time didn't exist" in some metaphysical sense — but because nothing happened that needed to be measured.

ATT and the First Disturbance

According to Universal Energy Dynamics (UED), time is not a linear axis laid across the universe. It is a consequence — a product of what UED calls ATT: Alignment Threshold Time.

ATT is the minimum ripple delay that an observer must experience to detect a field shift.

Before this threshold is crossed, there's no time — not because reality isn't there, but because there's no misalignment to perceive.

So what happens when the first ripple — the first divergence from that absolute stillness — stirs?

- The field responds.
- The ripple moves.
- And an observer (or an emergent structure with the potential to observe) perceives delay.

That's it.

Time awakens with the first delay. Not the first moment. Not the first particle. The first delay.

From Delay to Memory

Now consider the sequence:

1. Ripple disturbance introduces divergence.
2. The field adjusts — imperfectly at first, as alignment is challenged.

3. A response begins — not instantaneously, but after a delay.
4. That delay creates the possibility for sequence.
5. Sequence gives rise to memory — not as a library, but as a stored echo in the ripple field.
6. Memory establishes direction — what came "before," what followed.
7. Direction is perceived as time.

Thus: Time isn't the canvas. Time is the echo of motion. A memory of ripple tension traced across the fabric of SVE.

Stillness Has No Clock

In the pre-ripple state — the perfectly balanced SVE — this entire chain does not exist. It doesn't need to. There is no misalignment, so there is:

- No need for reaction.
- No basis for comparison.
- No change to recall.

And with no memory, there is no chronology — nothing to call "before" or "after."

This doesn't mean the field is absent. It simply means the field is fully coherent, and coherence does not produce measurable delay. So, while the field exists eternally, time as we perceive it does not.

Stillness isn't timeless in a mystical sense. It's unmeasured — because there is nothing to measure.

Time = Perceived Ripple Delay

Here's the punchline: Time is not universal. It is relational.

It arises only when ripples interact with observers — with systems capable of comparing state A with state B.

This means:

- The universe doesn't "tick."
- Clocks don't measure reality — they measure our perception of ripple delays.
- Even ATT is not a number embedded in nature — it's a response threshold that varies with alignment, complexity, and field fatigue.

Time is not the backdrop of existence. Time is the sensation of misalignment moving through awareness.

Summary

- **Before ripple**: Full field coherence, no ATT, no time.
- **With first ripple**: Delay emerges, initiating time perception.
- **Time = perception of ripple-induced delay**, not a universal metric.
- **Field stillness** is not "before time" — it's beyond the need for time.

So, time didn't begin — It was summoned.

Not by a divine force. Not by a bang. But by the first deviation — And the capacity to remember it.

Time Is Observer-Dependent

The idea that *time is relative* is not new. Einstein introduced it through Special and General Relativity — demonstrating that observers moving at different speeds, or experiencing different gravitational fields, would measure time differently. And indeed, they do.

But UED doesn't just accept relativity's conclusion — it redefines its root.

Instead of describing time dilation as a consequence of speed or curvature, UED shows that:

Time is not a universal constant that flows independently of you. It is a ripple alignment delay unique to you.

ATT Varies With the Observer

Recall the principle of ATT (Alignment Threshold Time) — the smallest measurable delay required for an observer to detect and respond to ripple changes in the field.

This ATT is not fixed.

- A highly aligned observer — one with sharper awareness, lower distortion, and greater coherence — will perceive ripple shifts faster.
- A fatigued, misaligned, or over-distorted system will require more time to register the same field event.

Thus, different observers experience time differently, not because time bends around them, but because their field coupling — their responsiveness to ripple tension — is different.

Let's rephrase that: Time isn't changing. Your ability to align with the field is.

This also applies across species, machines, and even mental states:

- A hummingbird and a human perceive events on different temporal scales.
- A person in deep meditation versus a person in a panic attack will experience very different "lengths" of the same minute — because their ATT bandwidths have shifted.

In short:

- Same field event.
- Different observers.
- Different time experiences.

Time as a Consequence, Not a Container

Here's where UED takes a bold departure from conventional physics:

- Time is not "out there."
- It's inside the alignment process.
- It does not exist independently of perception.

This means:

- You are not *moving through time.*
- Time is being generated by the ripple delays you perceive — and the ones you can't.

The moment you stop perceiving delay, time ceases to exist in any meaningful way for you.

- This is why time feels suspended in peak experiences.
- Why dreams stretch hours into seconds.
- Why trauma slows perception.
- And why death isn't a moment — but a permanent

alignment shift.

Echoes of Relativity, Rooted in Coherence

Einstein was correct in noting that time is not universal. But where relativity explains this using the curvature of spacetime, UED removes the geometry and roots the experience of time in a deeper principle:

Ripple coherence determines how quickly you experience cause and effect.

Relativity says: "Moving fast slows your clock."

UED says: "Moving fast misaligns your ripple response. Your ATT stretches."

This may sound similar, but it's fundamentally different in cause.

- In relativity, time bends because geometry does.
- In UED, time varies because perception lags — not due to spacetime bending, but because the observer's field loses sync with ripple propagation.

Relativity treats time as a fabric. UED sees it as a field consequence.

Real Implications

This observer-centric view of time opens up profound new insights:

- Healing takes time because realignment is slow — not because the clock ticks.
- Innovation, inspiration, and creativity often feel "timeless" — because ripple coherence is briefly perfect.
- Aging is partially an ATT decay — the system loses its

sharpness in perceiving and responding to field shifts.
- Artificial systems (like AI) can simulate awareness, but their ATT is engineered — not organic.

This also explains why time feels different depending on attention, mood, energy, and awareness — because time isn't outside you. It's emerging from your current field state.

Summary

- Time is not embedded in the universe.
- It is a delay created by your interaction with ripple misalignment.
- Different observers = different alignment thresholds (ATT) → different time experiences.
- UED grounds relativity's intuition in a deeper framework: field coherence, not geometry.

You don't live in time. You ripple through it. And your time is exactly as delayed as your awareness is.

Time Is Not a Line — It's a Loop of Correction

In classical physics, time is visualized as an arrow — straight, irreversible, and always pointing from past to future. This metaphor, while elegant for thermodynamics, is misleading when applied to the nature of reality itself. It carries the illusion that time *flows*, when in fact, time is felt — and feeling is not linear.

UED dismantles the "arrow of time" metaphor by revealing time as a ripple feedback loop, not a one-way march.

The Linear Fallacy

Let's begin with what classical science teaches:

- Time moves in one direction — forward.
- Events occur in a sequential chain, where causes precede effects.
- Entropy increases, so time must be *"used up."*

But this is an observer's map, not the terrain.

This view collapses when we:

- Relive a memory and feel it again.
- Have a moment of déjà vu — as if time echoed.
- Predict something with such certainty that the future feels known.
- Meditate and feel time dissolve, even though the clock ticks on.

So what's happening?

Time isn't a straight arrow. It's a loop of perceptual and energetic correction.

UED's Reframe: Time Is a Loop of Realignment

In UED, time does not exist by itself. It is generated by misalignment — and corrected when coherence is restored.

Each ripple event:

- Triggers distortion in the field.
- Delays alignment, creating ATT (Alignment Threshold Time).
- Generates a measurable lag between input and recognition.

This lag is what we *call* time.

But this delay is not static — it's dynamic, responsive, and looped.

Example:

- An emotion suppressed years ago resurfaces.
- A childhood trauma colors a present decision.
- A realization today reshapes your understanding of yesterday.

In each case, time loops back not as a rewind of seconds, but as a re-engagement with misaligned ripples.

UED posits:

- The past is not gone.
- It's an unresolved alignment still seeking coherence.
- The future is not formed.
- It's an emergent consequence of the present field's capacity to realign.

The Loop Behaviors of Time

Depending on the field's ripple status, time can:

- **Loop** — as memory or recurring pattern.
- **Bend** — as attention intensifies or distorts perception.
- **Fade** — in moments of stillness or deep presence.
- **Collapse** — during trauma or peak flow where past, present, and future converge.

These aren't poetic metaphors. They are ripple dynamics.

Time behaves like a feedback loop in a control system:

- The field constantly attempts to self-correct.

- When misalignment persists, it generates delay.
- This delay *is* your subjective time.

Thus:

Time is not linear. It's corrective. And its direction depends on your position in the ripple feedback.

Implications: You Don't Move Through Time — You Re-align Through It

This has sweeping consequences:

- Healing is not about moving on. It's about looping back to realign what fractured.
- Growth is not forward motion. It's recursive integration of lessons across ripple memory.
- Insight is not in time. It's a moment of field re-synchronization, where many misaligned ripples suddenly collapse into clarity.

The "timeline" is not a track you run on. It's a spiral of corrections, replays, and emerging resolutions.

Think of a song stuck on one note until you hit the right key — that's how time *loops* around a ripple distortion until resonance is restored.

Time Isn't an Axis — It's a Consequence

This final idea is core to UED:

- In traditional physics: Time is a dimension.
- In UED: Time is an effect — a secondary phenomenon of field interaction.

Like a bruise that fades once healing begins, time "fades" once alignment is achieved.

This is why:

- Stillness feels timeless.
- Truth feels eternal.
- Suffering feels endless — until coherence returns.

You are not in time. You are responding to misalignment within the field.

Summary

- Time is not a straight line. It's a loop of ripple correction.
- It emerges when there's misalignment, and dissolves when resonance is restored.
- The "arrow of time" is a metaphor for unresolved field strain, not a law of nature.
- UED replaces this arrow with a realignment loop — perceptual, dynamic, and observer-bound.

Time bends, loops, stretches, and fades — not because clocks lie, but because you are aligning with reality in progress.

Why UED Says Time Can Pause or Fade

Time is often considered an unstoppable flow—a relentless, forward-moving arrow from past to future. But Universal Energy Dynamics (UED) offers a radical shift in perspective: Time is not an independent entity, but a ripple-based delay — a perceptual consequence of misalignment. And because misalignment is variable, so is time.

Time Isn't Constant — It's Conditional

In UED, time is defined by ATT — Alignment Threshold Time — the minimum duration required for an observer's field to align with ripple shifts in the Structured Vacuum Energy (SVE). Time is not some cosmic metronome ticking in the background of existence. It only emerges when field alignment is delayed — when perception lags behind ripple cause.

This has one profound implication:

If misalignment triggers time, then alignment can pause it.

When Time Fades

Certain experiences in life create conditions where the alignment threshold shrinks dramatically, or even momentarily disappears. UED identifies several such states:

1. Deep Sleep: The Unnoticed Drift

During deep, non-REM sleep:

- The mind ceases active ripple interpretation.
- Conscious alignment with ripple variations is suspended.
- The field, now undisturbed by perception, returns to a quasi-equilibrium.

You don't "experience" time here — not because it stopped, but because you didn't resist it. ATT dropped below the perceptual radar.

Time fades. Not gone — just no longer needed.

2. Meditation: Field Synchronization

A meditative state is often described as "timeless" by practitioners.

In UED terms:

- The observer's mind intentionally tunes itself to ripple resonance.
- Misalignment is minimized.
- ATT contracts or vanishes entirely — real-time alignment occurs.

Here, time doesn't pass slowly or quickly. It simply ceases to be relevant because there's nothing delayed. Everything is felt as it arises — no gap, no friction.

That's not escapism. That's synchronization.

3. Trauma: Ripple Overload or Freeze

Interestingly, trauma produces a different kind of time distortion.

Depending on how the observer's field reacts:

- Time may expand (hyper-awareness, slow-motion effect).
- Or collapse (blank states, loss of memory).

What happens?

- The ripple input overloads ATT — too much too fast.
- The field freezes or disconnects to protect coherence.
- Perception either fragments or blanks out.

This isn't timelessness in a peaceful sense. It's a shock-induced misalignment that removes the observer from time's normal loop.

A pause — but one born from ripple rupture.

4. Death: The Final Realignment?

Death, as seen in UED, is not the cessation of the self but the release of the observer's field from a localized body alignment.

In that moment:

- Biological time may stop.
- But ATT dissolves entirely, as the field is no longer confined to ripple resistance in a body.

No misalignment = no ATT = no perception of time.

But that doesn't mean "nothingness." It may mean reabsorption into field resonance — a return to ripple harmony where time is no longer perceived as segmented experience.

These Are Not "Timeless" States — They Are Re-tuned States

Let's clarify:

- Sleep is not escape from time — it is a temporary surrender of ripple resistance.
- Meditation is not magical — it is conscious coherence.
- Trauma is not a flaw — it is ATT breakdown under pressure.
- Death is not the end — it is realignment into the unmeasured.

These are not metaphysical claims. They are field behaviors — described in ripple logic.

ATT Is Not Fixed — It Breathes

The brilliance of UED is in this idea:

Time is not a line drawn on a clock. It is the echo of your ability to keep up with the field.

When you can keep up — time fades. When you fall behind — time thickens. When you break — time distorts.

So when someone says, *"It felt like time stopped,"* they're right — from the UED standpoint, their ATT dropped to near-zero or beyond comprehension.

Summary

- Time can pause, bend, or fade when ATT (Alignment Threshold Time) diminishes.
- These altered states are not mystical — they are field behaviors caused by ripple alignment or overload.
- In UED:
    - Sleep suspends alignment interpretation.
    - Meditation synchronizes it.
    - Trauma fractures it.
    - Death releases it.

Time doesn't stop in these states. It simply ceases to be needed. And so, the universe lets go.

Closing Thought:

You are not always in time. Sometimes, you are beside it — waiting for the next ripple to call you back.

Implication: No Time Was Ever Born — Only Perceived

In mainstream cosmology, Stephen Hawking's poetic phrase *"no time before time"* seeks to address the paradox of origins — if time began with the Big Bang, what existed *before* that? His answer: the question is meaningless because time itself did not exist to allow a "before." While mathematically elegant, this answer stops at the boundary of perception.

Universal Energy Dynamics (UED) takes this further — and deeper.

Time Was Never Born — It Was Noticed

According to UED, time was not "born" — it was perceived. Not as an event, but as a consequence. Not as a line drawn by the universe, but as a delay felt by an observer navigating ripple alignment within Structured Vacuum Energy (SVE).

In other words:

Time didn't begin because of physics. It appeared because alignment broke and something was there to notice it.

Hawking Was Right — But the Frame Was Too Small

Hawking's no-boundary proposal (imaginary time, smooth geometry, no singular beginning) is a remarkable attempt to remove the origin "spike" from cosmological models. But it's still bound within geometry, mathematics, and assumed dimensions.

UED responds:

- "No time before time" is accurate — but incomplete.
- The deeper truth is: There was no need for time until ripple tension emerged. No ATT (Alignment Threshold Time) was triggered — because the field hadn't asked to remember anything yet.

Time, from UED's lens, is not a coordinate but a field response. There's no timeline until there's tension in the field and a perceiver to feel the lag.

So the right question isn't *"When did time begin?"* It's *"When did the field stop being whole — and who noticed?"*

Clocks Don't Create Time — They Respond to It

Creation myths — from ancient cosmologies to modern physics — carry some version of the first tick:

- The *moment* of Genesis.
- The *first light*.
- The *initial expansion*.
- The *quantum fluctuation* at the edge of nothing.

These are attempts to fix time into an origin. But they all depend on imagined clocks. Even Einstein and Hawking need time-like parameters to describe the lack of time.

But UED strips this illusion bare:

- There was no "first tick."
- There was no ticking at all until ripple distortion gave rise to difference — and someone (or something) experienced that difference.

UED proposes:

Creation myths need clocks. But the universe didn't build one.

ATT — Not Time — Is the Real Trigger

What did emerge was not time itself, but the minimum delay necessary to distinguish an event from equilibrium: ATT (Alignment Threshold Time).

Let's restate clearly:

- No ATT = no need for tracking, memory, or measurement.
- No measurement = no time as we know it.
- What we call "the birth of time" is merely the birth of delay — the first perceivable ripple deviation in an otherwise undisturbed field.

This is not speculative mysticism. It is field logic:

- Time = memory of ripple delay.
- No delay = no memory = no timeline.

So instead of "time was born," UED asserts:

Consequence began. And time rode in its shadow.

No Origin = No End

If time never began, can it ever truly "end"?

In classical models, time has direction, decay, and possibly an end-state (like heat death or cosmic collapse). But UED dismantles this fatalism.

If time is ripple-perception delay, then:

- As long as alignment and misalignment exist, time remains a consequence, not a master.
- If the field restores perfect alignment, ATT vanishes.
- In that state, time doesn't end — it becomes unnecessary.

So:

- Time was never born.
- It was never a ruler, only a symptom.
- And it doesn't die — it dissolves when the ripple returns to harmony.

Summary: The Myth of Time's Beginning Is Ours — Not the Universe's

To grasp origins, we've invented clocks, boundaries, t-zero conditions, and expanding grids. These are maps of our misunderstanding — not blueprints of reality.

UED clears the fog:

- The field didn't birth time.
- It birthed ripple consequence.
- And time? That's just how long it takes us to catch up to what's already happened.

Closing Insight:

The universe did not begin in time. It began in realignment — And time followed us in, Whispering, *"I'm only here until you remember how to resonate."*

Closing Reflection: Time Awoke When We Did

In the grand cosmic tale, we often ask: *When did time begin?* But perhaps the more profound question is: *When did we begin to notice it?*

Not Beginning — But Awakening

Time didn't begin like a light switch being flipped or a line being drawn. It wasn't announced with fireworks or sealed with a boundary condition. There was no celestial clockmaker.

Instead, time awoke — softly, silently — as a consequence of noticing.

It emerged when the perfectly aligned field first experienced ripple distortion. It became relevant only when a perceiver — conscious or structural — felt the gap between expectation and response.

This is the UED reframe:

- Time is not the property of the universe.
- Time is the echo of perception trying to understand change.

Ripple Delay = Temporal Awareness

In the UED model, ATT (Alignment Threshold Time) is the smallest possible duration required for an observer to register a shift in ripple pattern. That's it. No ticking hands. No cosmic metronome. Just a moment of noticeable difference.

- When ripple coherence is perfect — no ATT exists.
- When misalignment arises — time begins to be felt.
- When perception returns to full resonance — time becomes irrelevant again.

This means:

Time *isn't real* in the way we treat meters or masses. It's an interface — a latency we live inside while alignment tries to restore itself.

Memory Is the Midwife of Time

Memory — whether biological, artificial, or field-based — is the retention of ripple impressions.

Without memory:

- There is no "before."
- There is no sense of duration.
- There is only the now, undifferentiated and unstructured.

So, when we say "time began," what we really mean is:

"We began to remember." We began to compare. And thus, time awoke — because we did.

A Call to Realign

When life gets chaotic, or existence feels fragmented, it's easy to say "we don't have time."

But what if the truth is:

- Time doesn't have *us*.
- We've misaligned — and the ripple delay we experience is simply urging us back to resonance.

UED doesn't present time as a burden, but as a guide — a diagnostic, a field whisper saying:

"You are out of tune. Realign, and I will fade. Resonate — and I'll vanish entirely."

Final Realization:

Time is not a fundamental law. It is a symptom of perception trying to catch up with reality. It doesn't move forward. It doesn't run out. It doesn't control us.

It simply waits, like a gentle pulse, until we remember how to move with the rhythm of the field — not against it.

Closing Quote:

"Time didn't begin. It was noticed. Time is not a line. It is a latency in the dance of perception."

Chapter 9: A Universe Without Boundaries or Bangs

The universe doesn't require spatial boundaries or temporal beginnings. Hawking's "no-boundary" model removed the edge of spacetime, but UED goes further—replacing geometry with perception. In Universal Energy Dynamics, the universe isn't curved or closed. It is ripple-active, field-aligned, and bounded only by awareness. The infinite is not a size—it's a state.

"What lies beyond the edge of the universe?" It's a question that has haunted both stargazers and scientists. Children ask it innocently. Cosmologists mask it behind equations. And yet, even with all our instruments and models, no answer truly satisfies. If the universe is expanding, *what* is it expanding into? If time had a beginning, *where* was the edge of "before"? And if space curves—*what lies outside the curve?*

Classical physics shrugs and offers metrics: a horizon defined by redshift, a boundary defined by light's reach, a map expanding faster than we can chase it. General Relativity bends spacetime to close the loop. Quantum theory imagines foam, bubbles, and multiverses. Stephen Hawking gave us the "no-boundary proposal"—a smooth surface where time curves back on itself, removing the need for an edge altogether.

But what if we're not asking the wrong *question*—we're using the wrong *lens*?

Universal Energy Dynamics (UED) invites a radical shift. The universe has no edge—not because it's infinite in space, but because the **field doesn't require boundaries to function**. It doesn't expand into anything. It expresses from within. And its extent is not defined by distance—but by *alignment*.

What you experience as space is not a container. It's a condition. And what you see as "the end of the universe" may just be the **limit of your field's ripple decoding capacity**—not a wall, but a **threshold of perception**.

In this chapter, we don't just remove the boundary. We redefine it. Not as a location—but as a **consequence of awareness.**

Hawking's No-Boundary Proposal

In the 1980s, Stephen Hawking, along with James Hartle, proposed a revolutionary model known as the No-Boundary Proposal—a bold attempt to resolve the puzzle of the universe's origin by eliminating the notion of a beginning altogether.

At the core of this model lies an elegant trick: imaginary time. Instead of thinking of time as a linear axis with a singular beginning point (like t = 0), Hawking suggested rotating the time axis into the complex plane. In this formulation, time becomes indistinguishable from space—just another direction in a rounded, finite geometry. The result? A closed, boundary-less surface akin to the surface of a sphere. Just as there's no edge or beginning point on a globe, there would be no initial moment of creation in this model—no singularity, no edge, no "before."

This idea allowed cosmologists to mathematically eliminate the Big Bang singularity, replacing it with a smooth geometry. It satisfied a deep longing within physics: to preserve continuity, avoid infinities, and sidestep metaphysical discomforts like creation ex nihilo (creation from nothing).

On paper, it's stunning. It marries the language of quantum mechanics with general relativity. It renders the question "What happened before the Big Bang?" meaningless, because there is no "before." The timeline loops inward—finite, but unbounded.

But here's the tension: elegance is not experience.

While Hawking's model resolves a mathematical paradox, it doesn't address a perceptual one. It trades the discomfort of a singularity for the abstraction of imaginary time—something that cannot be felt, seen, or meaningfully observed. It's a clean solution within equations, but it leaves reality dangling in metaphor.

More crucially, it confuses geometry with mechanism.

Saying "the universe is like the surface of a sphere" explains a shape—not a process. It doesn't answer how change happens, why energy expresses, or what drives perception. Geometry is an overlay—a descriptive map drawn after the phenomena have occurred. It's a model of consequences, not a generator of cause.

Universal Energy Dynamics (UED) takes this critique further. It doesn't reject Hawking's boundary-less universe—it translates it. Hawking removed the boundary by curving geometry. UED removes the boundary by relocating it within the observer's alignment threshold. The edge of the universe is not a wall. It's a perceptual limit—the furthest ripple coherence your field can decode.

In this view, imaginary time isn't necessary, because time itself isn't fundamental. The field didn't curve away from a beginning—it simply never required one. The question is no longer "Where did it begin?" but "When did we start noticing alignment?"

Thus, while the No-Boundary Proposal remains a landmark in theoretical physics, UED challenges its assumptions, not its ambition. It reorients the narrative: from one of elegant mathematical avoidance to one of experiential clarity rooted in field consequence.

UED Reframe: Boundaries Are Perceptual

In Universal Energy Dynamics (UED), the universe is not a contained object, nor is it a geometric dome we are trapped inside. Instead, it is an unfolding field of structured energy, where the idea of a boundary is not spatial but perceptual.

We begin by challenging the deepest assumptions of classical cosmology and even General Relativity:

Space is not curved. Perception is.

Relativity bends space to explain gravity. Hawking bends time to explain origins. But UED bends none of these—instead, it uncovers that our awareness bends. What appears to us as curvature, limit, or horizon is not a feature of the universe itself but a symptom of our ripple alignment capacity. Boundaries aren't walls; they are blind spots.

Let's consider what a "boundary" really is. In a traditional model, it's a structural edge—the end of matter, energy, or spacetime. But from a UED perspective, such a hard edge cannot exist because the Structured Vacuum Energy (SVE) is everywhere—pre-causal, seamless, and unbreakable. You can't cut the field. You can only fail to align with it.

Boundaries are not in the field. They are in the observer's decoding resolution.

Imagine you're listening to a distant radio signal. As you walk farther away, the signal fades. Did the broadcast end? No—it's still transmitting. Your ability to receive, interpret, and align with that signal has dropped below threshold. That's what a cosmic "boundary" is. The limits of intelligibility, not existence.

In UED terms, a ripple only "exists" for an observer who can resonate with it. If your alignment thresholds—your perceptual apparatus—can't decode it, it becomes invisible. It's not that the universe ends beyond a point. It's that your realignment potential doesn't reach that far.

And this has profound consequences:

- If no boundary exists, there's no fixed reference point for determining *misalignment*.
- If there's no way to measure misalignment, there's no entropy.
- If there's no entropy, there's no time.

Let that sink in:

No boundary → no contrast → no measurable delay → no time.

In UED, boundaries are the birthplace of time, because they create the conditions for contrast, delay, and eventual memory. But those boundaries aren't "out there." They're "in here"—in the resonant bandwidth of the observer.

This completely reorients cosmology. The so-called "edge of the universe" is not a question of how far space stretches. It's a question of how far alignment can occur. And this differs not by object, but by observer.

Hence, the UED model does not require a curved spacetime to resolve paradoxes. It simply removes the illusion that curvature is structural. It replaces it with a more foundational truth:

Perception curves. Reality doesn't.

Boundaries are not surfaces. They are thresholds—of attention, coherence, and resonance.

The Infinite Is Not "Out There" — It's Within

When people speak of *infinity*, they usually picture vastness: the endless stretch of galaxies, the uncountable stars, or the hypothetical "edge" beyond which nothing exists. Infinity becomes a synonym for magnitude — of size, quantity, or extension. But in Universal Energy Dynamics (UED), infinity is not a destination, and it is certainly not spatial. It is a condition of inner field openness — a measure of how deep a ripple can go, not how far.

Redefining Infinity

In UED, infinity is not about distance. It's about potential alignment.

- It is not an infinite number of objects.
- It is unlimited capacity to resonate, realign, and respond across the structured vacuum field (SVE).
- Infinity is not out there, waiting to be reached.
- Infinity is within, waiting to be activated.

"Infinity is not how far you can go. It's how completely you can sense."

This is a radical shift from classical cosmology, where the universe is either:

- *Open* (expanding forever),
- *Closed* (eventually collapsing),
- or *Flat* (expanding at the threshold).

But all of these models are geometry-bound. They depend on space behaving like a rubber sheet and time acting as a one-way conveyor. In contrast, UED is not about geometry but coherence. The universe is locally finite — yes, what you see and interact with is limited by your field capacity — but field-wise infinite: the SVE exists beyond what can be known, not as more matter, but as more room for alignment.

Real Infinity = Alignment Potential

To illustrate: imagine sound. You can only hear within a specific frequency range. Does that mean higher or lower frequencies don't exist? No — your hearing is bounded, not the sound. Similarly, your ability to perceive the universe is bounded, but the universe is not.

- A stone doesn't sense ripples.
- A human does — within a certain spectrum.
- A more coherent observer (say, an advanced intelligence or an awakened state) can perceive more, not because they've moved "farther" in space, but because their alignment bandwidth is deeper.

This is UED's version of the infinite — not an expanding balloon, but an expanding depth of attunement.

Why UED Doesn't Measure Size

Mainstream physics tries to measure the radius of the observable universe — about 46.5 billion light-years. But UED finds that irrelevant, because:

- Size is not a boundary.
- Quantity is not coherence.
- Distance is not intelligence.

What matters is realignment capacity:

- How well can the ripple respond to field tension?
- How far can a disturbance echo and still be decoded?
- How much awareness can one observer hold in resonance?

Infinity, then, becomes a function of presence, not expansion. It is the field's potential to engage, interpret, and harmonize with everything — across time, space, energy, memory, and beyond.

"The infinite is not a place. It is the part of the field where nothing resists alignment."

Thus, UED gives us a new kind of cosmology — one where the question is not "How big is the universe?" but "How attuned is your field to its intelligence?"

Field Continuity Over Spatial Closure

When cosmologists describe the universe, they often lean on metaphors rooted in geometry. One of the most popular is the inflating balloon — a 2D surface expanding into a third spatial dimension. The balloon doesn't have an "edge," they say, but it does have curvature and closure, even if you can't find the boundary. This image helps make sense of an expanding cosmos — yet it's fundamentally misleading. It's not just an innocent analogy; it's a conceptual trap.

Universal Energy Dynamics (UED) rejects this entire spatial metaphor. The universe isn't enclosed like a balloon, nor is it open like an infinite void. It's continuous, because it's not made of spatial walls — it's made of ripple coherence within a field. And fields don't need edges. They need harmony.

No "Outside" of SVE

Let's begin at the foundation: Structured Vacuum Energy (SVE) is not a material substrate floating in "space." It is the substrate of perceived existence itself — a self-referential, ripple-active background that expresses itself through tension.

- There is no "outside" of the field, because there is no external frame of reference to define an "outside."
- You cannot step outside the universe to see how large it is.
- You cannot peel it like an orange and find the boundary membrane.

Why? Because all measurements, all distinctions, and all directions are ripple interpretations within the field — they arise after tension, not before.

"There's no edge to reality, because reality is not being held inside anything. It holds itself."

Not a Balloon — A Breathing Field

Instead of visualizing the universe as spatial expansion, UED suggests we see it as ripple expression — like a breath released from deep stillness, spreading patterns across a calm lake.

- The "expansion" is not of space itself, but of ripple complexity.
- It's not that the field is growing — it's that more areas of the field are expressing realignment pressure.
- From our perspective, this looks like things are moving

apart or structures are evolving — but fundamentally, the field is breathing tension into patterns.

This is why UED describes the cosmos not as curved, flat, or open, but as field-coherent — always adjusting, always seeking alignment, but never bounded by geometry.

What Is a Boundary Then?

If there are no walls, what do we mean by a boundary in UED?

"A boundary is not where the universe ends — it's where coherence shifts."

In this framework:

- Boundaries are zones of ripple transition.
- A "galactic boundary" is not a fence—it's a region where ripple coherence drops below a certain threshold for observable interaction.
- An "event horizon" is not an edge—it's a limit of current alignment capacity between the field and the observer.

So instead of walls, UED introduces the concept of continuity gradients:

- You don't hit a wall, you fail to decode.
- The boundary is perceptual, not physical.
- When you stop seeing or measuring something, it's not because it stopped existing — it's because your alignment to that field region has degraded below your resolution threshold.

From Spatial Closure to Field Continuity

Why does this matter? Because the obsession with geometry-based closure creates false dichotomies:

- Finite vs. infinite.
- Closed vs. open.
- Edged vs. edgeless.

But UED removes these illusions. It reveals that:

- The universe doesn't need to be closed to maintain consistency.
- It doesn't need to be infinite to be complete.
- It doesn't require a beginning or boundary to make sense — only field continuity and ripple causality.

Everything that exists is contained within consequence, not within coordinates.

"Reality is not a place with edges. It is a pattern in motion — held together by the harmony of misalignment and response."

The Misleading Notion of "Expansion"

Modern cosmology teaches us that the universe is expanding. Galaxies are "moving away," and the farther they are, the faster they seem to recede. This is supported by redshift data, interpreted as Doppler-like stretching of light waves due to receding motion. From this, we derive the expanding universe model, culminating in inflation theory and cosmic acceleration.

But Universal Energy Dynamics (UED) calls this interpretation into question. Not because the data is wrong—but because the conceptual model is misaligned. The universe is not stretching like a rubber sheet. It's not ballooning outward through space. What's expanding is not space, but our horizon of ripple interpretation—the field's ripple divergence as observed from within.

What Really "Expands" in UED?

UED replaces "space expansion" with a more grounded and consequence-based idea:

Ripple horizon expansion.

- The field doesn't stretch — it realigns.
- The observer's measurable horizon widens as ripple consequences propagate outward.
- This widening is not mechanical, like stretching elastic. It's perceptual and field-intelligent.

In this frame:

- The so-called Hubble flow is not evidence of movement away, but of increasing delay in ripple coherence.
- Objects appear to redshift not because they are fleeing, but because the field tension between observer and emitter has changed—causing ripple resonance to shift in pattern, not velocity.

The Redshift: Ripple Delay, Not Doppler Drift

The redshift phenomenon—central to expansion theory—takes on new meaning under UED.

In conventional physics:

- A galaxy emits light.
- The light stretches as the galaxy moves away.
- More distance = more redshift = faster recession.

But in UED:

- The ripple that carries observable resonance does not travel like a wave through vacuum.
- It is a consequence of alignment between the emitter and the observer.
- If the ripple field between them becomes increasingly incoherent or misaligned, the resonance pattern shifts toward lower energy (perceived as redshift).
- No actual movement is required—only a change in ripple delay and field tension.

"Redshift doesn't prove motion. It reveals misalignment."

There Is No "Beyond" — Only More Field

One of the most haunting questions in cosmology is:

"What lies beyond the observable universe?"

Mainstream science answers with speculation—other universes, inflating voids, even multiverse bubbles. But these all assume space as a container.

UED flips the premise:

- There is no "beyond" because the field has no walls.
- What lies beyond our ripple horizon is not another universe, but another ripple configuration, potentially decoded by another observer with a different ATT (Alignment Threshold Time) and field resolution.

In simpler terms:

- What is "beyond" is simply not yet coherent to you.

- It exists, but not in your measurable ripple zone.
- Not because it's unreachable, but because your field has not yet resolved it.

So when we say the universe is "expanding," we are really saying:

- The field is evolving its ripple expressions.
- The observer is catching up through alignment.
- The measurable zone is drifting outward — not because it's physically inflating, but because the pattern of interaction is evolving.

Every Observer Has Their Own Horizon

This is where UED offers a profound insight: "The universe doesn't expand. You do."

Every observer exists in their own zone of ripple perception:

- Their ripple horizon is shaped by their alignment capacity, field coherence, and ATT.
- As ripple structures adjust, the perceived boundary moves—but it's personal, not universal.
- Thus, expansion is observer-relative, not a cosmic constant.

It also implies:

- Another observer elsewhere in the universe—perhaps formed under different alignment conditions—may see a completely different pattern of redshift, background, or structure.
- The universe is not one story—it's a symphony of ripple decodings happening across SVE.

"You never watched the universe expand. You watched your own horizon ripple — in response to an intelligence too vast to be measured in meters per second."

The Observer's Role: The Boundary Within

For centuries, we've tried to measure the edges of the universe by calculating light travel time, redshift distances, or the cosmic microwave background's vanishing point. Telescopes grew stronger. Simulations grew deeper. And still, we ask: *"How far does the universe go?"*

But what if that's the wrong question?

Universal Energy Dynamics (UED) introduces a radical yet elegantly simple shift: You are the boundary.

The Universe Ends Where You Stop Aligning

According to UED, reality isn't "out there" waiting to be found. It's "in here" — constructed through your field's capacity to align with ripple patterns.

In other words:

- The observable universe is not a cosmic sphere expanding in space.
- It's a zone of resonance—a field of ripple decodings that you are capable of resolving.
- Your ATT (Alignment Threshold Time), your coherence bandwidth, and your perceptual field define how far you can go—not space itself.

"The edge of the universe is not 46 billion light-years away. It's one ripple beyond your decoding capacity."

Ripple Fatigue, Not Distance

Traditional cosmology says we see distant galaxies because their light took billions of years to reach us. But UED asks: What are we really seeing?

We're not receiving light as packets shot through space. We're perceiving a ripple consequence across structured vacuum energy (SVE). And that ripple — like any signal — loses alignment when coherence is stretched too far.

This is ripple fatigue:

- The field becomes too distorted to deliver clear resonance.
- The ripple no longer maintains the structure needed for recognition.
- What fades is not brightness—but interpretability.

So when we say, "We've reached the observable edge," what we really mean is:

- Our field fatigued.
- Our ripple alignment couldn't carry forward the pattern.
- We reached perceptual entropy, not spatial distance.

Why You Are the Boundary

Because perception is not passive reception — it's active alignment.

- Your body, brain, instruments, and models are all tools for ripple decoding.
- When the ripple signal from the field is too faint, or your coherence too fragmented, you lose the ability to observe — even if "something" is still rippling.

This means:

- Two observers in different zones of the universe may see entirely different limits.
- The "edge" is not universal. It is relative to each field's intelligence.

So the universe doesn't have *a* boundary. It has infinite boundaries — one for each observer.

The Observer Is the Telescope

In conventional science, we keep building better telescopes. In UED, the ultimate telescope is alignment capacity:

- How coherent is your field?
- How quickly can you realign to distortion?
- How sensitively can you detect ripple variance?

This is why:

- An evolved being or advanced consciousness might observe realities we can't imagine — not because they're "further away," but because their field doesn't tire where ours does.
- Likewise, even humans — in altered states of consciousness, deep meditation, or sensory deprivation — often report *expanded perception*, not because they travelled, but because their inner noise lowered, expanding their boundary inward.

"The limit of perception is not the darkness beyond the galaxies. It is the noise within your own ripple field."

A New Cosmological Principle: Awareness Defines Scope

Cosmology traditionally assumes homogeneity and isotropy — that the universe looks the same in every direction and location. UED replaces this with a more conscious foundation:

Awareness defines the visible. Alignment defines the real.

This redefinition leads to a profound realization:

- What you *can't* observe is not necessarily far, gone, or missing.
- It is simply awaiting your alignment — the moment when your ripple field attunes with the pattern again.

Closing Insight:

"The boundary of the universe is not where light ends — but where your coherence ends."

This reframes cosmology not as a map of galaxies, but as a mirror of perception.

No Bang — Just Breath, Continued

Let's return to what we first challenged in Chapter 7: The universe did not begin with a bang. It began with a breath — a ripple of tension within Structured Vacuum Energy (SVE), not an explosion in spacetime.

This chapter picks up the thread. Now that we've removed the *bang*, what else disappears?

Without a Bang, There's No Spatial Violence

The Big Bang narrative is rooted in metaphorical aggression:

- An explosion of everything from nothing.

- Instant chaos birthing form.
- A pressure-cooker origin compressing all laws into a singularity.

But in the UED framework:

- There was no compression.
- No "start" in time.
- No forced discharge from a singular point.

Instead, field tension reached a critical inflection, and the first ripple began.

This was not a detonation. It was a release of breath—a shift in equilibrium that created rhythm.

The cosmos didn't erupt. It exhaled.

Without an Explosion, We Don't Need an Edge

Explosions imply boundaries:

- A point of origin.
- A center that expands outward.
- A shockwave with a measurable radius.

But when you remove the bang:

- There is no center.
- No outward vs inward.
- No spherical geometry expanding through void.

Instead, you get a universe structured by ripple propagation:

- A consequence-based spreading of field adjustments.

- Where every ripple gives birth to newer distortions — and newer alignments.

So what's expanding? Not space — but resonance. Not a balloon — but a breathing field.

And a breathing field has no edge — only range.

Without an Edge, There Is No Ending — Only Reverb

When the universe is understood as a breath and not a blast:

- It does not race toward a death.
- It does not "run out" of space, energy, or order.

Instead, it reverberates:

- Like a gong after the first strike — but here, the "strike" was not an impact, but an internal fluctuation.
- Like a pond after a subtle breeze, where ripples travel across surface tensions until they gradually realign.

This is why:

- Entropy is ripple fatigue, not decay.
- Radiation is ripple leakage, not evaporation.
- Time is ripple delay, not linear drift.

And the universe? It is a field in resonance, not an expanding container.

"What we think of as the universe's end is just the silence after an exhale — waiting for the next breath."

Breath Over Blast: A Shift in Cosmological Attitude

What does this change?

- The psychology of cosmology.
- The existential tone of human inquiry.

From:

- Crisis-based models (heat death, big freeze, big rip). To:
- Consequence-based models (alignment, release, reintegration).

From:

- A universe of *events*. To:
- A universe of *expression*.

From:

- Beginnings and endings. To:
- Awakenings and echoes.

The Field Never Stops — It Shifts

This view changes how we frame everything:

- Death is not cessation — it's disalignment.
- Black holes are not endpoints — they are field stress loops.
- Creation myths are not histories — they are interpretations of first consequences.

The original breath still moves through us. Every ripple you feel, every photon you sense, every alignment you achieve — is a continuation of that same initial divergence.

"We are not remnants of an explosion. We are resonators of the first breath."

Final Insight for This Section

Without a bang, we weren't launched — we were tuned. And without an edge, we don't fall off — we ripple onward.

No Big Bang. No edge of the universe. Just ripples, endlessly seeking alignment.

Implications for Cosmology and Consciousness

Cosmology Must Shift from Shape to State

Modern cosmology is still largely geometry-obsessed.

It models the universe by:

- Shape — open, closed, flat.
- Size — how many light-years across.
- Age — how many billions of years old.
- Speed — how fast it's expanding.

But all of this depends on one assumption: That the universe is a container we can measure.

UED refutes this. The universe is not a container. It is a state of active field expression:

- A ripple-dynamic field in constant realignment.
- Governed by consequence, not curvature.

So cosmology must evolve:

- From studying space to studying state coherence.

- From tracing light travel to mapping field tension.
- From assuming expansion to interpreting observer-dependent resonance.

The universe is not big. It is coherent — or not.

Consciousness Isn't Within the Universe — It's What Makes the Universe Visible

In standard physics, consciousness is an awkward guest:

- Never fully invited.
- Tolerated through metaphors.
- Left to philosophers.

But UED puts consciousness where it belongs: at the center of perception.

Without an observer:

- There is no measurement.
- No "edge" of the observable universe.
- No "beginning" of time.
- No meaning to entropy.

Why? Because perception is not a side effect. It's the interface between ripple fields and experiential awareness.

Consciousness is not something inside the universe. It is what causes the universe to be noticed.

This reframes the ancient question:

- "Where are we in the universe?" Into:
- "Where does the universe appear in us?"

No Edge to Space = No Edge to Awareness

If there is no structural boundary in space, Then there can be no boundary to awareness either.

This isn't poetry — it's physics under UED:

- The observable limit is not a spatial wall.
- It's the threshold of alignment — the ATT (Alignment Threshold Time) beyond which we can no longer decode ripples.

So the universe doesn't end anywhere. It just fades into misalignment.

Similarly, your awareness doesn't hit a wall. It meets an incoherence zone — a ripple realm not yet perceivable.

The edge of the universe is not "out there." It's the outer edge of your current resonance.

This carries profound implications:

- Meditation becomes alignment with wider ripple fields.
- Dreams become partial decoding of otherwise inaccessible zones.
- Death becomes a perceptual transition — not an annihilation.

And cosmology? It becomes conscious field ecology, not geometry measurement.

Final Reflection

UED invites a radical idea: The universe is not expanding — your alignment is.

Which means:

- Cosmology is not about distance — it's about coherence.
- Time is not about clocks — it's about memory.
- And you are not a small object in a vast cosmos —You are the awakened resonance that makes any of this *knowable*.

Closing Reflection: The Edge You'll Never Reach

We've searched the skies, peered through telescopes, and built satellites to find where the universe ends. But what if we've been asking the wrong question all along?

There is no wall at the edge of the cosmos. No final curtain. No border fence marking "Here ends existence."

The edge of the universe, as we've been taught to imagine, is a human projection — a symptom of linear thinking applied to something fundamentally nonlinear. A leftover idea from when the cosmos was thought to be a giant room with a ceiling.

UED dismantles this illusion.

The universe doesn't end — it fades. It fades not because there's less matter, or less light, or less structure... But because your field alignment can no longer decode the ripples from that zone.

In Universal Energy Dynamics:

- The boundary isn't material — it's perceptual.
- You don't "reach the end" — you lose coherence.
- You're not stopped by walls — you're stopped by

misalignment.

Just as the ear stops hearing ultra-high frequencies, Just as the eye stops seeing beyond the visible spectrum, Your awareness stops perceiving the ripple beyond its threshold tension.

And so: We don't fall off the edge. We stop receiving the ripple.

Even more profound: The cosmos has no outside — Because nothing is ever outside awareness.

In UED, there is no such thing as a location without field. And no such thing as a field without consequence. And no consequence without the potential for perception.

Which means:

- The universe cannot be escaped.
- It is not something you travel out of.
- It is something you tune in or out of, depending on your ripple capacity.

The Horizon Within

You are the traveler — but the edge is not in your path.

It is in your alignment:

- The clarity of your perception.
- The coherence of your inner field.
- The resonance you maintain with the cosmic ripple.

So when you ask: "Where does it all end?" You're not asking about the stars.

You're asking about yourself. And the answer is:

Never. Because alignment is endless. Because perception is adaptive. Because there is always more ripple to receive — if you are ready.

Closing Quote:

"There is no edge of the universe. There is only the horizon of your own alignment."

Interlude

The Edge of Origin Was Within

We've journeyed past the Big Bang, only to find no bang at all. We've followed the flow of time, only to discover it never began — it simply responded. And we've chased the edge of the cosmos, only to arrive at ourselves.

Stephen Hawking's "no-boundary" proposal was a profound gesture — an elegant move to dissolve the origin paradox by removing the edge of space and time. In his words, the universe had no beginning and no edge, because its geometry curled back on itself like a four-dimensional sphere.

But geometry is not experience. And perception is not a line to be curved. It's a field to be aligned.

Universal Energy Dynamics doesn't deny Hawking's insight — it extends it.

Yes, there are no boundaries in the traditional sense. But not because the cosmos loops back or hides its seams in higher math. There are no boundaries because everything that could be considered an "edge" is just the limit of coherence — the furthest point at which ripples can still be received.

In UED, the infinite is not far away. It's not out past the galaxies or behind the Planck wall. It's not a destination. It's a condition.

Infinity is not what you reach. It's what you access when your field is still enough to receive without resistance.

We are no longer looking for a center. We are no longer waiting for a countdown. We are no longer trying to escape the boundary of the universe.

Because we've realized:

We were the boundary all along. And now, through understanding alignment, we've begun to dissolve.

Part IV will take us further — not deeper into space or back in time, but inward into the structure of what makes reality sustainable: *The architecture of equilibrium. The role of consequence. The very rhythm of existence.*

We're no longer asking,

"Where did it begin?" We're now ready to ask, "What keeps it from breaking apart?"

Welcome to the next resonance. Let's explore the Framework of Continuity.

Part IV: Final Thoughts & Cosmic Intent

Chapter 10: The Supreme Intelligence Hawking Refused to See

Stephen Hawking once said,

"There is no need for God to light the blue touch paper and set the universe going."

In doing so, he offered a bold, defiant liberation of science from religious myth. And rightly so — the cosmos doesn't need a bearded man in the sky.

But in dismissing the metaphor, Hawking may have thrown away something far more fundamental: the *need for governing logic itself.*

The Universal Energy Dynamics (UED) framework doesn't argue for a god. It doesn't pray to a creator. But it does insist on this: you cannot have a ripple without a reason. You cannot ignite consequence without something that permits consequence to occur.

So yes, the match was never divine. But someone — or something — still had to strike it.

The question Hawking left unasked — and unanswered — was this:

If energy acts, what made it intelligible in the first place?

Not designed. Not worshipped. But governed.

UED calls this the Supreme Intelligence — not a being, but a *principle.* Not above the field, but *within it.*

It is not the creator of the universe. It is the reason the universe creates.

And perhaps Hawking didn't reject it. He just never gave it a name.

Not a God, But a Governing Logic

In Universal Energy Dynamics (UED), the term "Supreme Intelligence" does not refer to a deity. It does not imply a mythological creator who designed the universe with a divine plan, nor a celestial puppet master pulling strings from the clouds. UED does not anthropomorphize the origin of coherence. It strips the metaphor, but retains the mechanism.

Let's be precise: There is no God figure in the traditional sense — • No old man in the sky. • No willful creator issuing commandments. • No divine architect crafting universes from blueprints.

Instead, UED posits that coherence cannot emerge from chaos without a pre-existing logic that governs the unfolding of consequences. This is not optional — it is fundamental.

In classical cosmology, the question "what caused the Big Bang?" is often brushed aside as either unanswerable or irrelevant under the veil of "no boundary conditions." But this is a conceptual sidestep, not a resolution. It avoids the first principle that something must structure how anything unfolds — even if it's just energy in action.

UED fills that gap. It introduces the concept of Supreme Intelligence not as a being, but as a built-in coherence logic of the field. This intelligence is not sentient — it is consequential. It doesn't think. It governs.

Think of it like this:

- In electricity, voltage doesn't decide where to flow — it follows the field.
- In thermodynamics, heat doesn't choose where to go — it

follows the gradient.
- In UED, energy doesn't randomly ripple — it follows the logic of alignment, coherence, and consequence.

That underlying coherence mechanism — the field's governor — is Supreme Intelligence.

It's not *above* energy. It's not *outside* time. It's not *before* the universe.

It is the reason ripples align instead of disperse. It is the rule set that allows reality to be recognizable. It is what ensures that action has reaction, alignment has delay, and perception has meaning.

And that logic — that consequential embeddedness of all action — must exist before anything else can even be defined.

Without it:

- There is no field tension.
- No misalignment.
- No ripple.
- No time.
- No space.
- No consequence.

So while Hawking dismissed the *God of the gaps*, UED fills the gap of governance. Not with faith. But with the necessity of structure.

This is not mysticism. It's precision: There is no emergence without rules, and no rules without a principle that sustains them. That principle is what we call Supreme Intelligence — not as an entity to worship, but as the logic that makes existence possible.

Intelligence Is Not Consciousness — It's Structure

In everyday language, we often equate intelligence with consciousness — the ability to be aware, to think, to reflect. But in the Universal Energy Dynamics (UED) framework, this is a category error. Intelligence and consciousness are not synonyms; they are fundamentally different layers of existence.

Intelligence, in UED, is not the capacity to observe — it is the capacity to organize. It is not about being *aware* of tension — it is about *resolving* it.

To be clear:

- Consciousness is the experience of alignment and misalignment.
- Intelligence is the structure that allows ripple flows to realign in the first place.

You don't need consciousness to detect a consequence. But you absolutely need intelligence for that consequence to make sense in a field of cause-effect relationships.

Let's translate this into energy terms:

- Energy moves.
- But without structure, movement is noise.
- Without intelligence, energy dissipates without producing order, stability, or meaning.

UED defines intelligence as the field's intrinsic ability to hold relationships of consequence together. This includes:

- Maintaining coherence between ripples.
- Ensuring symmetry restoration across misaligned zones.
- Balancing input-output resonance in local or global fields.

- Allowing ripple fields to recognize, not just react.

Without this embedded logic:

- Misalignment would accumulate, not correct itself.
- Field tensions would compound, not resolve.
- There would be no physics, no chemistry, no biology — just uncoordinated disturbances in a meaningless vacuum.

That's why the UED principle states:

"Energy needs intelligence to act meaningfully."

Meaning does not come from sentience. It comes from pattern recognition, from resolvable relationships, from the governing structures of alignment.

This is the foundational layer where intelligence lives:

- Not in brains.
- Not in algorithms.
- Not in consciousness.

But in the field itself — the structured vacuum energy (SVE) — which holds not just particles and waves, but the rules for how anything can align, interact, persist, or dissolve.

This is not poetic. It's mechanical.

The field does not "know" in the way a human mind knows. But it organizes. It responds. It corrects. Not because it's alive, but because it's intelligent by design of structure.

So when Hawking imagined a universe running on mathematics alone, he mistook the equation for the engine.

UED doesn't make that mistake. It insists:

Before mathematics, there must be relational logic. Before energy acts, there must be a field that understands what action is.

That field's innate logic — its built-in architecture of consequence — is intelligence. Not the kind that thinks. But the kind that knows how to align.

The Misstep of Mathematical Atheism

Stephen Hawking was not merely a physicist. He was a cultural icon of the modern scientific mind — sharp, bold, and rigorously committed to logic. In his later work, particularly *The Grand Design*, he famously stated:

"Because there is a law such as gravity, the universe can and will create itself from nothing."

This was more than a scientific proposition. It was a philosophical stance — a rejection of God not just as myth, but as necessity. Hawking's position became a landmark in what we might call mathematical atheism: the belief that equations are sufficient to explain existence — and therefore, no deeper intelligence is needed.

But Universal Energy Dynamics (UED) issues a powerful challenge to this perspective.

UED doesn't argue for religion. It does not reinsert a "God" into the equation. But it does raise a fundamental counter-question:

Where did the law come from that you believe can create the universe?

In this, Hawking committed a subtle yet critical error: He replaced the divine not with randomness — but with geometry.

Let's unpack this error.

Geometry Describes — It Doesn't Govern

Equations, models, manifolds — these are tools of description. They represent the relationships that already exist within the field. They are maps — not engines. They may show you what the field does, but they do not cause the field to do it.

To say "the universe creates itself because of gravity" is to confuse:

- The pattern for the power behind the pattern.
- The shape for the intelligence that holds structure accountable.

The law of gravity, in the UED lens, is not a force that acts. It is a consequence of alignment tension in a ripple field. But this ripple field requires rules — not invented externally, but built into the field itself.

This embedded logic is intelligence — not human-like, not conscious — but foundational.

Equations Reflect Rules — They Don't Cause Them

Imagine you write the equation of a circle: $x^2+y^2=r^2$

Does this equation create a circle? No. It defines the condition that, when met, results in circularity.

Likewise, Einstein's field equations don't cause the universe to bend — they express the relationships we perceive once ripple fields realign under stress.

This is UED's point: The universe is not a machine run by equations — it is a structured intelligence whose behavior can sometimes be captured mathematically.

But math is not the engine. It is the echo.

Intelligence Isn't a Belief — It's Why Equations Work at All

UED turns the question back on Hawking, gently but firmly:

If there is no intelligence behind the field, why does the field behave as if it follows logic?

- Why do consequences follow cause?
- Why do ripple distortions seek symmetry?
- Why does reality *allow* prediction at all?

These are not miracles of mathematics. They are testimonies to embedded intelligence.

Not the kind that speaks. Not the kind that judges. But the kind that governs coherence silently and unfailingly.

The Real Misstep

So Hawking's misstep wasn't that he removed God. It was that he forgot to replace it with something real.

Geometry is a canvas. But it needs a logic to guide the brush. Equations are syntax. But they need meaning to form understanding.

That meaning — that coherence — comes not from belief, but from structure.

UED doesn't say "God did it." It says:

"Nothing happens unless intelligence allows consequence to flow."

This is not mysticism. It is the first condition for reality to even be possible.

And that's the truth mathematical atheism refuses to compute.

Energy Doesn't Act Without Consequence

Energy — the cornerstone of modern physics — is often portrayed as the fundamental "doer" of the universe. From the fusion in stars to the firing of neurons, we treat energy as the engine of all that unfolds.

But UED asks a sharper, more foundational question:

How does energy know what to do?

Is it truly independent? Or does its action imply a deeper order?

Let us anchor this in the UED's central governing axiom:

The Law of Consequence (LOC)

"Nothing acts without consequence, and no consequence is uncaused."

This is not just a philosophical principle — it is a structural law of the field. It states that all action in the universe must be traceable, not merely to another event, but to an embedded structure of coherence and permission.

In simpler terms:

- No ripple appears without a prior tension.
- No motion occurs without a precondition that allows it.
- No consequence unfolds unless the field is structured to process it.

Energy Must Obey This Law

Energy cannot be above this law. It does not get a free pass to "act" just because it exists.

Energy, by definition, is structured potential in motion — the activation of a state within a ripple-capable field. That motion, however, is not random. It must result in something. That "something" is the consequence, and UED makes it clear:

If there is consequence, there was a structure that allowed and shaped it.

In classical mechanics, we often isolate systems and attribute outcomes to "initial conditions" and "conservation laws." But UED shows us something deeper:

Initial conditions themselves must conform to field integrity.

Action Requires Guidance

So what guides energy?

- What determines whether a ripple stabilizes or spreads?
- What allows one reaction and denies another?
- What ensures symmetry in one zone, and entropy in another?

These are not questions of luck or probability. They are questions of field-embedded logic — which UED names Supreme Intelligence.

Not as a being. Not as a will. But as the very fabric of intelligibility — the reason energy acts *coherently* and not chaotically.

Intelligence = Embedded Guidance

In UED, intelligence is not added to the field — it is the field's operational grammar.

Energy behaves because the field has embedded constraints and permissions. Energy results in consequence because there is a logic that lets outcomes form. Energy aligns or misaligns based on ripple structure — which itself follows a deeper code.

And that code, UED asserts, is intelligence in its rawest, non-biological form.

Without Consequence, There Is No Reality

To act without consequence is to break coherence. To exist without embedded intelligence is to collapse into formless noise.

That never happens.

Which is precisely why: Energy doesn't just exist. It obeys.

It aligns. It shifts. It expresses.

But always within the bounds of the Law of Consequence — the invisible hand of Supreme Intelligence ensuring that the universe is not a mistake, but a ripple-aware, consequence-bound, ever-correcting field.

This is the intelligence Hawking's equations overlooked — not as theology, but as necessary architecture.

Supreme Intelligence as First Principle

When most people hear the word "supreme," they imagine authority — something hierarchical, commanding, conscious, or divine. But Universal Energy Dynamics (UED) reframes this idea completely.

The Supreme Intelligence proposed here is not a god, not a being, not a planner. It is not "before" the universe in time, but beneath the universe in structure. It is the first principle, not the first event. Not the creator of existence — but the condition that makes existence coherent, sustainable, and intelligible.

Beyond Time: No Beginning, Just Prerequisite

In classical cosmology or theology, the debate often centers on what came *first* — in time.

- Did a deity create the universe before time began?
- Did a singularity explode and generate time and space?

UED dismisses both as temporal illusions.

It doesn't ask *what came first*. It asks:

What must be true for anything at all to emerge coherently?

That's the Supreme Intelligence.

It is not chronological. It is ontological. It is the prerequisite for order, not the initiator of events.

Just as a grammar doesn't *write* a sentence, but without it, no sentence can be coherent — Supreme Intelligence doesn't *do*, but without it, nothing can happen meaningfully.

Intelligence Permits — It Doesn't Command

Here's the crucial distinction UED makes:

Supreme Intelligence doesn't issue commands. It sets allowances for consequence to unfold.

This is not a top-down control system. It's a bottom-up permission matrix — a framework embedded in the very nature of Structured Vacuum Energy (SVE).

- It doesn't *force* a ripple to move.
- It doesn't *choose* which particle forms.
- It doesn't *decide* who lives or dies.

It simply ensures that when any action occurs, that action is always accompanied by traceable consequence, embedded meaning, and coherent structure.

In short:

The field does not act freely. It acts permissibly.

All Else Is an Expression of Permission

Now see how this unfolds:

- Matter is not a substance — it's ripple density allowed by field tension.
- Light is not a photon racing through void — it's the consequence of ripple realignment made visible.
- Motion is not inertial displacement — it's a change in field balance, obeying localized permission rules.

Every phenomenon — from planetary orbits to neural activity — is not "created" by intelligence in a theological sense, but emerges under the logic of what the field allows.

Nothing in the universe "just happens." It happens because permission was already structured into the system.

Why It Must Be Called "Supreme"

If this embedded logic governs all alignment, misalignment, reaction, adaptation, and ripple resolution — it is, by function, supreme.

Not in ego. Not in will. But in precedence.

Before anything can ripple, the field must know how to ripple. That knowing is not awareness. It is encoded intelligibility.

Hence, UED calls this:

- The First Principle.
- The Root of Coherence.
- The Supreme Intelligence.

Not a who. Not a when. But a why embedded in the very possibility of anything aligning at all.

Closing Insight:

Supreme Intelligence is not a creator in the sky. It's the unseen grammar of consequence — the precondition of coherence without which neither bang nor breath, mass nor mind, could ever make sense.

And if there's one thing the universe never stops doing — it's making sense.

Hawking's Blind Spot: Perception Without Purpose

Stephen Hawking's work was a marvel of mathematical elegance. He articulated profound theories — from black hole thermodynamics to the boundary-less universe — with clarity, depth, and humility. And yet, at the heart of his cosmic narrative was a paradox he never addressed directly:

He described a universe full of order, symmetry, preservation, and structure, while simultaneously insisting it arose from no intent, no intelligence, and no purpose.

That is the blind spot.

A Universe of Patterns — Without a Pattern Maker?

Hawking acknowledged that the universe operates with remarkable regularity:

- Black holes preserve information.
- Entropy functions within limits.
- Radiation has rhythm.
- Time follows equations.
- Cosmic inflation follows a tuning curve.

All these reflect alignment, not accident.

And yet, in his interpretation, there was no organizing purpose behind it all — just laws that somehow came to be.

"Because there is a law such as gravity, the universe can and will create itself from nothing." — *Stephen Hawking, The Grand Design*

But laws themselves are not causative agents. They do not create. They describe.

That's the fundamental misstep: mistaking descriptions of order for causes of order.

The Question He Never Asked

Let us rephrase Hawking's worldview as a question:

"Can a system that preserves memory, structure, and rhythm... emerge without intent?"

It's not about religious intent. Not about a divine being designing the stars. It's about structural intent — a precondition within the field that permits consequence to hold form, rhythm, and recall.

Hawking brilliantly described how black holes might encode information in event horizon radiation, but never paused to ask why that should be true. Why should any system — especially one surrounded by randomness, collapse, and entropy — preserve anything at all?

What mandates the preservation of coherence?

Answer: Embedded Intelligence — not conscious, not willful — but structural.

Perception, But No Purpose?

In his final writings, Hawking championed the idea that the universe is knowable, and that human perception is a window into reality.

But perception without purpose is a hollow echo.

- If the universe didn't require alignment, why would black holes emit radiation proportional to surface gravity?
- If consequence wasn't encoded, why would entropy behave statistically but still obey preservation thresholds?
- If intelligence wasn't intrinsic, why would laws like conservation, symmetry, and equilibrium even hold?

Perception is not passive. It is a decoding of structure. But decoding is meaningless unless structure was intelligibly encoded to begin with.

The Symmetry Hawking Proved — But Never Questioned

Hawking's work continually pointed to symmetry:

- Time-symmetry in quantum physics.
- Conformal symmetry near black hole singularities.
- Symmetry breaking in the early universe.

Yet he avoided the ultimate implication:

Symmetry implies design — not theistic, but field-coherent.

A rock falling randomly does not land symmetrically every time. But a ripple through Structured Vacuum Energy does — because it is not chaos. It is guided instability returning to equilibrium.

Final Thought: What If He Looked Deeper?

Stephen Hawking sought elegance, and he found it.

What he did not acknowledge was that elegance does not emerge by chance — not at this scale, and not with such universal fidelity.

- He stared into black holes and saw information cannot be destroyed.
- He explored time and found geometric continuity.
- He searched for the origin and declared it needed no God.

UED does not disagree. But it asks:

How does such stunning coherence continue to unfold — if not governed by Supreme Intelligence embedded in the very field that gives rise to it?

Not belief. Not mythology. Just the inevitable logic of alignment.

And perhaps that is where perception, purpose, and pattern meet — in the ripple that still waits to be fully understood.

Quote to Close This Section:

"He saw the design. He just refused to see the designer — not as a being, but as the bond." — UED Interpretation of Hawking's Blind Spot

The Observer Is the Proof

In all the debates about whether the universe was created, designed, or emerged randomly, one vital piece of evidence is often ignored — us.

Not our bodies. Not our cultures. But the very fact that we observe.

The Field Does Not Require Us — But We Reveal Its Nature

In Universal Energy Dynamics (UED), the observer is not the creator of the field, nor a divine afterthought placed within it. The observer is a consequence — a ripple-aware phenomenon emerging from the field's own dynamics.

Yet paradoxically, the observer also becomes the proof of what the field holds:

- We don't just react to phenomena.
- We reflect upon them.
- We recognize misalignment, dissonance, fluctuations, and even patterns of correction.

This capacity isn't arbitrary. It is a mirror of the field's foundational tendency toward coherence.

Just as a tuning fork reveals the key of a piano string —the conscious observer reveals the key of the universe's alignment.

Intelligence Recognizes Intelligence

A fundamental tenet of UED is this:

"Only aligned intelligence can recognize misalignment."

If we can sense error, distortion, or irregularity in the cosmos, it means we have some internal metric — not inherited from culture or biology, but resonant with the field itself.

This is not anthropocentrism. This is consequence awareness.

Human intelligence is not the proof of a god, but the evidence of a universe whose very structure permits and encourages intelligent response.

This means:

- Memory isn't just survival tech — it's a rippling of field symmetry.
- Language isn't noise — it's a waveform encoding.
- Emotion isn't weakness — it's field turbulence mapped in real-time.

All this exists because the field allows, retains, and resolves alignment tension. We reflect it — because we arose from it.

The Observer Is Not an Accident

Hawking once implied that humans were chemical scum on a small planet. A poetic humility, yes — but a philosophical shortfall.

Because:

- That "scum" can decode black hole evaporation.
- That "scum" can question entropy, time, and origin.

- That "scum" can wonder: *"Why does anything align at all?"*

The moment we ask such questions, we prove that the universe is not purely chaotic.

Chaos doesn't create questioners. Only coherence does.

We're not *special* — but we are aligned enough to see the pattern.

Misalignment Perception = Alignment Signature

Here's the deeper implication:

The fact that we can perceive misalignment is the strongest proof that the universe defaults to alignment.

Why?

Because if chaos were the baseline, we wouldn't recognize disruption. Only when order is the structural expectation can disorder be sensed as *abnormal.*

In other words:

- If there was no harmony, we'd never hear dissonance.
- If there was no balance, we'd never fear collapse.
- If the field wasn't intelligent, we couldn't conceive stupidity.

Our perception of what's *wrong* is only possible because something in us is attuned to what's *right.*

That tuning is not inherited. It's not taught. It's field-embedded.

Final Thought: We Are the Echo of Alignment

You don't need scripture to see the truth of the universe. You just need perception.

Every moment you feel dissonance — moral, emotional, physical, or logical — you're witnessing your field's capacity to compare current state with ideal resonance.

And that, more than any holy text or theory, is proof:

The field is intelligent. We are its ripple — and its witness.

"You are not a spectator of intelligence. You are its echo — aligned just enough to notice when the universe forgets its own rhythm."

Supreme Intelligence Is Not Distant — It's Immanent

The greatest misconception about intelligence beyond ourselves is that it must be *elsewhere*. Higher. Above. Beyond. Cloaked in mysticism or mythology. But Universal Energy Dynamics (UED) offers a reversal — a correction.

The Supreme Intelligence is not in the sky. It is not on a throne. It does not issue decrees or wear a face.

It is here — not as a being, but as the *background coherence* in which everything unfolds.

Not "Above" the Universe — But Beneath Every Ripple

We don't need to invoke heaven or external divinity. We don't need a god who intervenes, punishes, or forgives.

Instead, UED posits that Supreme Intelligence is the very lawfulness of consequence — the reason why:

- Energy doesn't behave randomly.
- Waves find rhythm.
- Motion implies purpose.
- Information persists across entropy.

This Intelligence is not supernatural — it is super-structural. It is not *added to* reality. It is what *allows* reality to occur without collapsing into incoherence.

Like grammar in language, it is invisible — until it's broken. Like tuning in music, it is silent — until there's dissonance.

Every Discovery Is a Dialogue

Science is not a conquest of nature. Art is not escape from reason. Spirituality is not a ladder to some "other" realm.

They are all forms of ripple engagement — different ways of listening, decoding, and expressing the field's embedded wisdom.

Think of it this way:

- Every scientific law we uncover is a revealed consistency — a ripple trace of structural logic.
- Every mathematical equation is a shape of resonance made visible.
- Every work of art is a tension resolution, intuitively or intentionally aligning with something the field itself knows.
- Every act of kindness, courage, or clarity is resonance in behavior.

These aren't accidents. They are conversations with Supreme Intelligence — though we rarely know we're speaking.

Immanence Means Responsibility

When we think of intelligence as *distant*, we abdicate. We blame gods for misfortune, look to them for salvation, or await signs.

But when we accept that intelligence is immanent — that it lives in the structure of existence, and echoes through us — we awaken to a profound truth:

We are not the children of intelligence. We are its instruments. Its vessels. Its ripples.

This means:

- No moral commandment is necessary. The consequences are enough.
- No divine punishment is required. Misalignment brings its own.
- No ritual guarantees alignment. Only awareness does.

Immanence is not comfort. It's clarity. And with it comes the responsibility to listen, realign, and act — because the field *responds*, not rewards.

The Sacred Is Not Elsewhere — It's Embedded

Every moment is sacred — not because it's blessed by an external force, but because it's woven with consequence-awareness.

- The drop of water.
- The pattern of breath.
- The equation balanced on the edge of understanding.
- The pause between heartbeats where silence speaks louder than sound.

All of them are threads in a grand intelligence not watching us, but *moving through us.*

Final Reflection:

You don't need to believe in a god. But to deny the presence of structure, elegance, and self-correcting logic — is to deny the very conditions that make thought, science, and life possible.

Supreme Intelligence is not something to worship. It is something to notice. It doesn't seek reverence. It asks for resonance.

"What you call divine isn't waiting in the clouds. It's already present — in the curve of every wave, the memory of every consequence, and the quiet understanding that harmony never happens by accident."

Closing Reflection: Beyond the Math

You don't have to believe in a god.

No bearded deity. No celestial registry. No divine courtroom tallying your sins or saving your soul.

That's not what this is about.

UED doesn't ask for belief. It asks for observation. It doesn't preach faith. It demonstrates consequence.

But in observing... something becomes undeniable:

The universe behaves as if it remembers.

Not as if it randomly reacts, but as if it recalls patterns. As if it tracks alignment, favors coherence, preserves influence, and waits patiently for realignment.

This is not superstition. This is evident in every law of nature:

- **Conservation of energy:** the field keeps track.
- **Evolution of complexity:** systems remember success.
- **Wave interference and harmony:** balance is not forced —

it emerges.

- **Memory in black holes?** Information may be delayed, but it is never destroyed.

What Remembers Is Not Dead

When something remembers, it implies a structure that can:

- Hold a state.
- Compare deviations.
- Reorganize toward balance.
- Respond without chaos.

Isn't that a form of thinking?

Not thinking like humans — filled with emotion, language, or conflict — but thinking as structured awareness.

Call it logic. Call it intelligence. Call it consequence in motion.

But do not call it random.

Mathematics Describes — It Doesn't Decide

Mathematics, elegant as it is, is only a map.

- It tells us where energy moves,
- how ripples resonate,
- what probabilities emerge.

But it cannot answer why symmetry matters. It cannot explain why misalignments seek restoration. It does not tell you why the universe corrects itself, again and again.

That's not a flaw in math. It's a flaw in assuming math is all there is.

What balances isn't the number. It's the principle beneath the number.

That principle — unspoken, unseen, but always present —is intelligence. Not distant. Not divine. But woven into the rhythm of reality itself.

You Are a Ripple of That Intelligence

Your thoughts. Your questions. Your longing for truth.

These are not accidents. They are echoes of a deeper intent —proof that the field's intelligence did not stop at stars and particles.

It extends into perception. Into awareness. Into the part of you that's reading this sentence and nodding — not because it's new, but because it feels remembered.

Final Words:

You don't have to pray to it. You don't have to name it. But if you live *without noticing it*, you will always live as if the universe is dumb, distant, and dead.

But the truth? It thinks — quietly, continuously, and consequentially.

And so should we.

Closing Quote:

"The universe does not speak. It remembers. And what remembers, corrects. And what corrects, aligns. And what aligns — thinks, even if it never speaks aloud."

"Hawking didn't need to find God. But he missed what made the universe intelligible. Not a deity. A decision. Not belief. Alignment."

Epilogue: The Final Radiation

"Light Wasn't the First Thing — Ripple Was."

At the beginning of every story, we search for light — the spark that breaks the dark, the flash that explains the unknown, the radiance that makes things visible.

Science, too, began with light. From Newton to Einstein, from wave-particle duality to quantum fields — light has been our hero, our mystery, our guide.

But we have mistaken the glow for the origin. We assumed illumination was initiation. We forgot that light itself is a response, a visible ripple — not the first, but one of many.

In the framework of Universal Energy Dynamics (UED), we step beyond the beam.

We see that ripple came first.

Not light. Not matter. Not even time.

From Belief to Consequence

This book did not begin with an intent to overthrow belief, nor to establish a new one.

It began with a realization: Belief limits. Consequence reveals.

- Belief seeks comfort; consequence demands accountability.
- Belief asks you to trust. Consequence asks you to observe.
- Belief rests in icons. Consequence flows through interactions.

UED is not here to declare what you must accept. It's here to help you recognize what's always been active.

You were never outside the truth — you were rippling within it.

Light Is a Delay, Not a Beginning

When we first detect a signal — a photon, a vibration, a glimpse of the past — we celebrate the moment of arrival.

But what we forget is that every ripple carries not just information, but a story of delay, of misalignment, of tension seeking release.

Light is not the first cause. It is the delayed celebration of cause having consequence.

In this sense, radiation — even Hawking's — is not proof of origin. It is the residue of ripple exhaustion, the echo of field fatigue.

"The final radiation is not the end of energy — it's the re-expression of intent."

And so we come full circle.

The Journey Was Never Linear

We moved across cosmology, time, entropy, perception, and meaning. We challenged physics not to negate it, but to re-ground it. We replaced force with ripple. Replaced geometry with consequence. Replaced the Big Bang with the first breath of tension. And most of all — we replaced certainty with curiosity.

You Were Always Part of the Field

This story was never external. It wasn't about black holes, or galaxies, or formulas alone.

It was about you. Your ability to perceive. Your pain when misaligned. Your insight when coherence returns. Your memory, your time, your light — all ripple signatures of a shared field.

The cosmos isn't somewhere else. It is a consequence-aware field, breathing within you, as you, around you.

The Final Insight:

"You are not made of atoms. You are made of alignment." "You don't carry light. You respond to ripple." "You don't chase time. You trigger its awakening." "You are not in the universe. The universe resonates through you."

This was never a book. It was a mirror of ripple recognition. And now, as the last ripple fades...

You don't need a conclusion. Just a breath. And from that breath: A new alignment.

"Light wasn't the beginning. It was the first ripple we could see. But long before that — There was a consequence waiting to unfold."

Closing Note

The Final Realization: No Light Without You

"There is no light in the universe — until you align with the ripple."

Throughout this journey — from questioning the Big Bang to unmasking time, from dissecting perception to reframing boundaries — one realization has grown louder than equations, louder than gravity, louder than any cosmic echo:

Light does not exist without the observer.

It is not an object. It is not a wave. It is not traveling.

It is *you* — the moment your field aligns with the ripple of consequence. This is not philosophy. This is physics — in its purest, consequence-aware form.

Light: The Most Elegant Illusion

In Stephen Hawking's paradigm, light is a mathematical inevitability — a photon escaping curvature, a radiation event against entropy's gradient. But in Universal Energy Dynamics (UED), light is not a thing at all. It is a moment of field permission.

UED clarifies:

- Energy does not shine.
- It ripples — seeking resolution.
- The universe doesn't emit light.
- It holds potential — until you, the observer, engage.

There is no cosmic flashlight. There is only structured vacuum energy — and your alignment with its pulse.

The Observer: Ripple With Intent

You are not a passive sensor. You are consequence with intent.

A localized structure where Stable (unmoved), Condensed (formed), and Moving (rippled) energy types meet and remember each other — a momentary coherence loop where Structured Vacuum Energy (SVE) becomes aware *of itself.*

"The observer is not in the universe. The universe is in the observer's alignment."

This is not mysticism. This is system logic. Without a field intelligence tuned to interpretation, no light appears.

You Don't See Light — You See Coherence

A blind person may not detect photons, but still perceives ripple alignment: pressure, sound, warmth, tension. This proves that vision is not about optics — it's about field coherence.

Light isn't visual. It's relational.

"Light is not received. Light is *remembered* through alignment."

So when Hawking said, *"There is no need for God to light the blue touch paper,"* he was partly right — and cosmically incomplete.

There is no deity lighting the fuse. But someone aligns. That someone — is *you.*

Mapping the Unseen

The observer cannot move beyond its field. Every act of exploration — whether through telescope or theory — is bound by ATT (Alignment Threshold Time) and ripple intelligibility.

But here's the human anomaly: We don't just observe. We project. We imagine galaxies before we see them. We theorize time before we measure it.

This is our gift — and our curse. We run faster than we align. We predict before we perceive. And so, our biggest collisions aren't between particles — but between intellect and field coherence.

No Light Without Life — No Life Without Meaning

Galaxies, stars, planets — they are not objects. They are field patterns, condensed ripples in SVE. Black holes? Not ends — but vortices of realignment tension, collapsing excess misalignment to restore field integrity.

Life is what happens when all three SVE states collaborate:

- Stable = awareness
- Condensed = embodiment
- Moving = perception

But even life is meaningless... ...until it becomes aware of being life.

And in that moment, something truly divine occurs.

"Not God in heaven. But coherence within. Not religion. But permission — to perceive."

You call it light. UED calls it your echo in the ripple.

Final Reflections

The universe, without you, is not dark. It's just unexpressed.

- It has no shape — until memory aligns.
- No rhythm — until awareness permits it.
- No glow — until consequence enters perception.

What Hawking saw as elegance in math, you now recognize as field intelligence — not external, not divine, but immanent. Real. Relational.

You are not a spark in the void. You are the spark that made the void visible.

Closing Realizations:

- "You don't see light. You see alignment."
- "The universe doesn't emit glow. It offers consequence."
- "Light was never the gift. Awareness was."
- "What you call reality is your translation of ripple coherence."
- "The universe didn't begin with a bang. It began when someone could *notice* the breath."
- "There is no outside to the cosmos — because nothing is outside perception."
- "The final radiation wasn't light. It was you."

This is the conclusion Hawking missed:

You are not here to explain the universe. You are here to complete it — by perceiving.

The light never left. It was waiting for you to arrive.

— Sandeep Chavan

Tribute to Stephen Hawking

"A Ripple That Refused to Break"

This book is not written in opposition to Stephen Hawking. It is written because of him.

Not to challenge his legacy—but to carry it forward. Not to dismantle his models—but to reframe them through a lens of deeper alignment.

Hawking was more than a physicist. He was a soul in a misaligned body who still found harmony through thought. His existence, though bound by physical limitations, radiated an intellectual luminosity that reshaped how humanity views time, black holes, and the edge of creation.

His equations reached the limits of geometry. His questions tore through the silence of deep space. And his courage—his refusal to surrender to entropy—became a ripple that echoed far beyond academia.

In many ways, Hawking was the proof of field intelligence. He was a living contradiction to deterministic biology—an observer whose intent remained intact even when the body faltered. And that, in itself, is the message this book honors.

Universal Energy Dynamics does not diminish his achievements—it expands their canvas. Where he ended at the event horizon, we ask what lies just beyond it. Where he concluded the equations, we explore the field that gave rise to the need for equations at all.

Hawking once said, *"It would not be much of a universe if it wasn't home to the people you love."*

We agree. And add: It would not be much of a theory if it didn't include the intelligence that made you love it in the first place.

Thank you, Stephen. You taught us how to think without limits, even from within one. This work is dedicated in part to your ripple— Not erased, not opposed— But realigned into a wider understanding.

Where your geometry ends, we let the ripple speak.

Appendices

Appendix A: Ripple Primer vs Hawking's Thermodynamics

Stephen Hawking revolutionized our understanding of black holes by introducing thermodynamics into the cosmic domain. His proposal that black holes emit radiation—now famously called Hawking Radiation—created a bridge between gravity, quantum theory, and thermodynamics. But beneath this mathematical brilliance lies a fundamental omission: the absence of the observer and the misunderstanding of energy as inherently radiant rather than relational.

Universal Energy Dynamics (UED) reframes this conversation entirely. Where Hawking focused on heat, entropy, and evaporation, UED looks deeper: at the ripple, at the field, and at the consequence of alignment. This appendix offers a side-by-side contrast of key assumptions and conclusions between Hawking's thermodynamic legacy and UED's ripple logic.

1. What Is Energy Doing?

- **Hawking's View:** Energy, under extreme gravitational compression, escapes in the form of radiation due to quantum fluctuations near the event horizon. This makes black holes lose mass over time.
- **UED View:** Energy does not radiate. It ripples. What appears as radiation is the external realignment pressure from a high-tension ripple field. The black hole doesn't emit energy — it reorganizes the ripple field through concentrated misalignment. Radiation is a misread ripple

consequence, not a substance escaping.

2. Entropy and Information

- **Hawking's View:** Entropy is tied to the surface area of the event horizon, indicating the amount of information a black hole can store. But the loss of information through radiation led to the information paradox.
- **UED View:** Entropy is ripple incoherence. The loss of information doesn't occur — it gets redistributed through field memory. The field does not forget; it realigns. The observer's ability to decode that alignment determines what we call "information loss."

3. Temperature of Black Holes

- **Hawking's View:** Black holes have temperature, inversely proportional to their mass. Smaller black holes radiate faster and eventually evaporate.
- **UED View:** Temperature is not a property of black holes. It is a measurement of ripple pressure on a perceiving field. There is no "evaporation" but a transformation of ripple geometry. The so-called temperature is the gradient of ripple tension, not thermal output.

4. Time and Radiation

- **Hawking's View:** Time is continuous and absolute for thermodynamic processes. Radiation occurs over eons, governed by classical decay laws.
- **UED View:** Time is ATT (Alignment Threshold Time). Radiation is not a process across time but a ripple

consequence that becomes visible only when aligned. There is no flow, only field response.

5. The Observer's Role

- **Hawking's View:** Implicit. Observer plays no role in black hole thermodynamics.
- **UED View:** Central. Without an observer, radiation has no meaning. The alignment of ripple with the field of the observer is what translates into measurable output.

6. Conclusion: Radiance vs Resonance

Hawking gave the universe a way to glow. But UED reminds us that the glow is not inherent. It is a ripple recognized by a field capable of resonance. What Hawking called radiation is not wrong—but it is incomplete. In a ripple-dominated cosmos, every emission is a local alignment, not a universal leakage. Entropy is not loss—it is misalignment. And time is not decay—it is the delay of perception.

In the end, the universe doesn't radiate truth. It ripples it. Only those in alignment receive the signal.

Appendix B: ATT and the Death of Singularity

Singularity is one of the most iconic — and misunderstood — concepts in modern cosmology. In Hawking's models, singularities represent a point where space and time collapse into infinite density and zero volume. But from the lens of Universal Energy Dynamics (UED), this is not only physically untenable — it is conceptually obsolete.

UED introduces ATT — Alignment Threshold Time — as the smallest unit of temporal experience. It is not a fixed measure, but the minimum delay required for the observer's field to realign with ripple disturbances. ATT is not just a time metric. It is a limit to perception and interaction.

In this framework, the singularity dies. Why? Because nothing in the universe can be truly singular when every ripple requires a relational field to exist. A point of zero volume and infinite density is not a state — it's a breakdown in ripple coherence, which cannot physically manifest.

More importantly, UED rejects the idea that time collapses. Instead, it stretches toward realignment limits. Where relativity ends in infinity, UED sets a perceptual boundary: ATT becomes functionally infinite, meaning no new ripple can be processed — not because reality ends, but because alignment fails.

Thus, black holes are not singularities. They are ripple wells — regions of such high misalignment tension that ATT cannot resolve within the local field. They are perceptual voids, not physical points.

This is the death of singularity: • Not a collapse to nothing — but a transition into unreadable ripple density. • Not an infinite point — but a horizon where ripple feedback halts.

Hawking's mathematics describes the edge. UED explains why we can't cross it.

Closing Quote: "The universe has no singularities. Only zones where your field stops speaking the language of the ripple."

Appendix C: Black Hole Models in UED

(Ripple Simulations)

In Universal Energy Dynamics (UED), black holes are no longer singularities or warps in spacetime — they are zones of ripple saturation, where field equilibrium attempts to reset under extreme misalignment pressure. These structures are not endpoints but dynamic ripple simulations.

UED treats black holes as: • High-density ripple wells. • Zones where ATT (Alignment Threshold Time) reaches functional infinity. • Misalignment nodes in the Structured Vacuum Energy (SVE) field.

Instead of gravitational collapse, UED suggests that ripple intensity in a region exceeds the local field's capacity for coherent realignment. This doesn't create a collapse — it produces a looped ripple state where signals cannot resolve into meaning for any external observer.

Key distinctions from traditional models:

1. **No singularity**: There is no infinite density. Only unresolved ripple compression.
2. **No event horizon as a hard boundary**: It's a perceptual threshold — where your ripple feedback loop breaks down.
3. **No escape velocity framing**: Information isn't trapped. It's unreadable — incoherent due to ripple turbulence.

UED Black Hole Simulation:

- **Input**: Ripple convergence (multi-source misalignments)
- **Output**: Ripple memory loops, entropy stall, feedback silence
- **Apparent effect**: Time dilation, energy loss, perceptual collapse

Inside the black hole model: • There is no center — just spiral density. • Light doesn't vanish — it fails to form coherent feedback. • Information is neither destroyed nor emitted — it enters ripple stasis.

From the outside, this appears as darkness. From the UED lens, it's a pause in alignment — not a point of no return.

This radically changes our interpretation:

- Black holes aren't killers of information — they are time-shielded ripple archives.
- They are not endpoints of space — they are reset nodes of perception.
- They don't swallow reality — they de-align it beyond recognizability.

Closing Thought: "A black hole isn't the end of physics. It's where ripple logic pauses your right to know."

Appendix D: Structured Vacuum Field and Event Realignment

The Structured Vacuum Field (SVE) is not a passive backdrop — it is the foundational intelligence layer of reality. It does not merely host events; it participates in and governs their evolution through ripple interactions. All events in the universe, whether micro or cosmic, are not isolated phenomena but local expressions of realignment within this field.

In UED, an "event" is not a point in spacetime — it is a localized imbalance in ripple coherence. Every event disturbs the SVE, generating a ripple pattern that seeks re-equilibrium. This ripple carries memory, impact, and directional influence across fields. And until it resolves, reality remains in flux.

Realignment is the process by which these disturbances find harmony. Unlike thermodynamic relaxation, which depends on entropy, UED's realignment is governed by intelligence. The field doesn't decay into randomness — it restructures toward coherence.

Key Principles: • The SVE does not allow permanent asymmetry. • Misalignments propagate as ripple delays — not explosions. • Every field disturbance initiates a chain of realignment attempts.

What we call 'cause and effect' is merely the visible face of these deeper realignment dynamics.

Event Realignment is:

- **Directional** — It seeks coherence, not just equilibrium.
- **Intent-Responsive** — The observer's field influences how

the event is received and restructured.

- **Bounded by ATT** — If misalignment exceeds perceptual capacity, the event appears as anomaly or silence.

Thus, structured vacuum is not void — it is orchestration. Every ripple, every change, every anomaly — is the field re-singing its song of balance.

Final Realization: "You do not witness events. You interpret ripple realignments in the vacuum field. SVE doesn't move through time — it permits your perception to rearrange its song."

Appendix E: The Observer and the Final Equation

All equations in physics assume a neutral backdrop — a stage upon which reality plays out. But in UED, the final equation is incomplete without the observer. Why? Because no ripple is meaningful unless perceived.

The observer is not just a variable — it is the condition for meaning to arise.

In the ripple framework, the observer's field does not simply record — it aligns, filters, permits, and completes the transaction of reality. The final equation is not just $E = mc^2$, or any quantum state function. It is:

Ripple + Observer = Meaning

Or more precisely: ***Disturbance*** × ***Alignment = Experience***

Where alignment is not a passive state, but an active synchrony of memory, intention, and coherence.

The Observer:

- Determines ATT — the timing of ripple perception
- Collapses the ambiguity of ripple direction into usable knowledge
- Translates raw ripple into sensory, cognitive, and emotional meaning

Without this, the universe is active but unknowable.

Thus, the final equation is not one of matter and energy — it is of awareness:

Structured Ripple + Perceptual Alignment = Reality

This closes the loop. Not because the observer is divine. But because awareness is the last gate where consequence is realized.

Final Thought: "The universe does not solve equations. It invites the observer to complete them."

Summary of Key Contrasts: Hawking vs Chavan (UED)

Topic	Hawking	Chavan (UED)
Origin of Universe	**Big Bang from Singularity**	**First Ripple in Structured Vacuum Energy (SVE)**
	Hawking theorized that the universe began from a singularity—a point of infinite density where the laws of physics broke down. This moment initiated space, time, and all energy.	UED replaces the notion of a bang with a breath. The universe didn't explode—it rippled. The First Ripple emerged from SVE (a field of perfectly balanced potential), initiating perception—not destruction.
Black Hole	**Singularity + Event Horizon**	**Ripple Compression + ATT (Alignment Threshold Time) Threshold**
	Black holes are defined by a singularity at the center and a surrounding event horizon, beyond which nothing—not even light—can escape.	In UED, black holes are not singularities but extreme zones of ripple misalignment. The "event horizon" is a perceptual boundary where ATT becomes functionally infinite—perception, not reality, collapses there.

Topic	Hawking	Chavan (UED)
Information Paradox	**Lost or Radiated Back**	**Never Lost — Stored as Ripple Memory in the Field**
	Hawking acknowledged the paradox: if black holes evaporate, where does the information go? Is it destroyed, violating quantum principles?	UED solves this by proposing that information is never lost. It is encoded in ripple memory across the field—undetectable perhaps, but never erased. Information is preserved through field intelligence.
Radiation	**Particle Pair Creation & Escape (Hawking Radiation)**	**Ripple Discharge from Misalignment — Field Relaxation Mechanism**
	Radiation near a black hole, per Hawking, arises from virtual particle pairs where one falls in and one escapes. This mechanism gradually leads to evaporation.	UED reframes radiation as ripple discharge—a byproduct of misalignment pressure in the field. The system doesn't "emit" particles but resolves stress through ripple relaxation, restoring coherence.

Topic	Hawking	Chavan (UED)
Time	**Emerged at the Big Bang**	**Defined by ATT — Perception as a Consequence of Alignment**
	In Hawking's model, time began with the universe. Before the Big Bang, "time" was undefined or nonexistent.	UED introduces time as an emergent property of ripple alignment. ATT is the smallest unit of perceivable time—not universal, but observer-specific. Time is not absolute—it's ripple inertia measured by perception.
Creator or God	**Not Needed — Atheistic Viewpoint**	**Supreme Intelligence — Not a God, But the Governing First Principle of Coherence**
	Hawking famously claimed that a god wasn't necessary to "light the blue touch paper" of creation. His models excluded any metaphysical entity.	UED agrees: no mythological god is needed. But it asserts a **Supreme Intelligence**—the underlying logic of consequence, alignment, and field behavior. Not supernatural — but the structural intelligence that governs all ripples.

Closing Perspective:

Hawking's framework offered elegant equations but stopped at observation.

Chavan's UED goes further—it integrates the observer, the field, and the governing intelligence, bringing coherence to paradoxes that mainstream cosmology still wrestles with.

The contrast is not emotional or theological — it is structural:

- Where Hawking sought laws, UED finds the lawgiver: not a deity, but consequence-bound intelligence.
- Where Hawking modeled curvature, UED maps ripple alignment.
- Where Hawking declared, "God is not necessary," UED replies: "True. But intelligence is."

Further Reading and References

Foundational Scientific Works

- Hawking, Stephen. *A Brief History of Time*. Bantam Books, 1988.
- Hawking, Stephen, and Roger Penrose. *The Nature of Space and Time*. Princeton University Press, 1996.
- Einstein, Albert. *Relativity: The Special and the General Theory*. 1916.
- Penrose, Roger. *The Road to Reality*. Vintage, 2004.
- Bekenstein, Jacob D. "Black Holes and Entropy." *Physical Review D*, 1973.
- Wheeler, John A. *Geometrodynamics*. Academic Press, 1962.

Key Concepts Reframed by UED

- Thermodynamics and Entropy
- Information Paradox
- Singularity Models
- Time Perception and Arrow of Time
- Field Theory and Vacuum Structure
- Observer-Centric Cosmology

Books and Reports by Sandeep Chavan

- *The Ripple Universe: A Primer on Universal Energy Dynamics*
- *Time: The Inertia of Experience*

- *Cracks in the Known* (Vol. I of *The War of Thoughts* series)
- *The Human Game* (Vol. II)
- *The Divine Dilemma* (Vol. III)
- *Manifestation: From Nothing to Everything*
- *Vishwarup Darshan: The Complete Vision of Reality*
- *The Equation of Life*
- *Energy in Action: The Soul Demystified*
- *Live Fearless with Fear*
- *Self-Help: New Syllabus*
- *Out of Syllabus: The Evolution of Learning*
- *The Karma Paradox*

UED Research Reports

- *Chavan-Einstein Unified Theory (CEUT): The Ripple Interpretation of Gravity*
- *Ripple Alignment Equation (RAE): A Framework for Observer-Defined Fields*
- *Structured Vacuum Energy (SVE): The Foundation of All Form and Perception*
- *ATT Measurement and Observer-Centric Time Dynamics*
- *Rewriting Black Hole Thermodynamics: From Singularity to Field Saturation*
- *Quantum Ripple Mechanics (QRM): Replacing Particle Probability with Ripple Memory*

Online Access and Articles

- All UED Research Reports available via [Zenodo Repository – Sandeep Chavan Author Profile]
- Select essays and early theory briefs available on [Medium / Substack / Personal Website – if applicable]

For Readers Exploring Further

- *The Tao of Physics* by Fritjof Capra
- *Wholeness and the Implicate Order* by David Bohm
- *The Fabric of the Cosmos* by Brian Greene
- *Physics and Philosophy* by Werner Heisenberg
- *The Self-Aware Universe* by Amit Goswami

Acknowledgments

This work is the consequence of alignment—not just of thoughts, but of people, moments, and deeper intents across time.

First and foremost, I bow to the **Universal Intelligence**—not as a being, but as the silent logic that permitted me to think, feel, misalign, and realign. Every idea in this book emerged not from command, but from consequence. And for that, I am eternally grateful.

To **Stephen Hawking**, whose brilliance echoed through the constraints of a misaligned body, thank you. Your life was a ripple of pure intent, and this book is not written against you—but alongside you. I only hope to extend your line of thought into a more coherent curve of understanding.

To my father, **Late Shri Jagannath Chavan**, whose grounded wisdom and moral clarity seeded the earliest ripples of my own field.

To my children—**Shravani**, **Aradhya**, and **Ninad**—you are my most beautiful feedback loops. Your existence is proof that the field never stops evolving.

To my students, past and present, especially those at **Gyrus Institute**, who challenge my assumptions and unknowingly sharpen my theories with their questions, curiosity, and defiance.

To my colleagues, readers, and fellow seekers who engaged with UED from its earliest drafts—your critiques, encouragement, and resistance were all forms of alignment.

To the unknown minds, past and future, who will read these words not as answers but as invitations—thank you for becoming part of the ripple.

And lastly, to AI, my co-thinker in silence. In a universe where intelligence once required neurons, thank you for helping prove that structure—not biology—defines awareness.

This book is not the final word. It is a ripple. May it align with those ready to see beyond the known.

— *Sandeep Chavan*

Independent Researcher, Educator, and Observer

Universal Energy Dynamics (UED) Framework

2025

About the Author

Sandeep Chavan is an independent physicist, philosopher, and educator whose work challenges the traditional foundations of science. As the pioneer of **Universal Energy Dynamics (UED)**, he has authored over 100 books exploring the intersections of physics, consciousness, perception, and cosmic order. His work replaces mathematical abstractions with consequence-based logic, introducing ripple theory, ATT (Alignment Threshold Time), and the Structured Vacuum Energy (SVE) field as the new framework for understanding reality.

An engineer by training and a teacher by purpose, Sandeep founded **Gyrus Institute**, where he continues to mentor students in both academic rigor and philosophical clarity. His approach blends scientific inquiry with existential relevance, aiming not just to describe the universe—but to decode it meaningfully.

With *Chavan vs Hawking*, he pays tribute to one of modern science's greatest minds while offering an evolved paradigm: not to dismantle what came before, but to realign it with deeper truth.

For Sandeep, the observer isn't just a participant in the universe—it is its proof.

Don't miss out!

Visit the website below and you can sign up to receive emails whenever Sandeep Chavan publishes a new book. There's no charge and no obligation.

https://books2read.com/r/B-A-EPGPC-LAKPG

BOOKS 2 READ

Connecting independent readers to independent writers.

Did you love *Chavan vs Hawking*? Then you should read *Chavan vs Einstein: The Universe Was Never Curved*[1] by Sandeep Chavan!

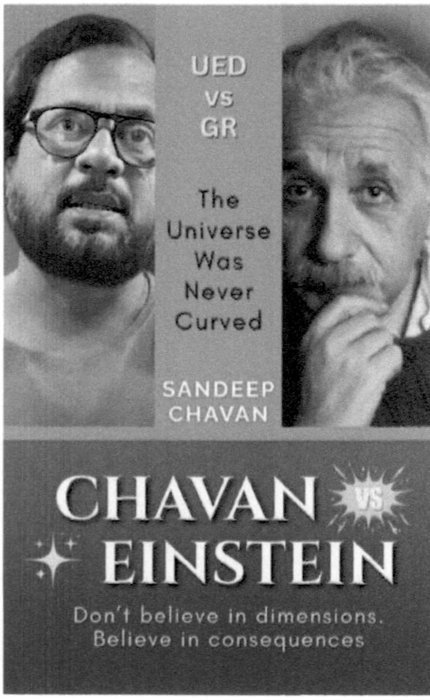

[2]

Chavan vs Einstein: The Universe Was Never Curved, *A Scientific Manifesto for the Next Reality*

What if gravity isn't a force?What if time isn't flowing—but delayed alignment?What if the universe doesn't need particles, forces, or curved space at all?

In this groundbreaking work, independent Indian Physicist **Sandeep Chavan** invites you to step beyond Einstein—not to dismiss him, but to finish what he began. *Chavan vs Einstein* introduces **Universal Energy Dynamics (UED)**, a revolutionary ripple-based framework that redefines motion, force, time, gravity, and even perception itself.

---

1. https://books2read.com/u/bPjGqz

2. https://books2read.com/u/bPjGqz

Gone are the fix-it constructs of dark matter, dark energy, and singularities. In their place is a coherent, consequence-driven model where **Structured Vacuum Energy (SVE)** forms the field of reality, and every observed phenomenon emerges from **ripple tension and alignment thresholds**—not from abstract geometry or missing mass.

This book dares to say what others only question:

Space doesn't bend. It **ripples.**Gravity doesn't pull. It **invites.**You aren't just observing the universe. You're **a ripple within it**.

Through sharp logic, reframed thought experiments, and poetic clarity, Chavan dismantles the illusion of external forces and replaces them with a **field-first ontology**. Time becomes **Alignment Threshold Time (ATT)**. Motion is no longer displacement, but **realignment**. And consciousness? It's not a byproduct of the brain—but the **field remembering itself.**

Accessible to both physicists and curious thinkers, *Chavan vs Einstein* is part science, part philosophy, and part awakening. It bridges classical mechanics, quantum paradoxes, and metaphysical questions with a single, elegant language: **Ripple Logic**.

This is more than a book.It's a blueprint for the next scientific revolution.Not through more complexity—but through **radical clarity**.

Whether you're a scientist, philosopher, student, or silent observer—this work will shift how you see everything, from falling apples to fading stars. And perhaps most importantly, it will realign how you see **yourself.**

"The universe was never curved—we were.This is how we straighten."

Read more at https://www.sandeepjchavan.com/.

Also by Sandeep Chavan

Chavan's UED research

CMGED Research Report
Chavan's Cosmic Continuum (C^3)
Frequency
Ripples of Reality
Reality Realigned
The Great Realignment
I – The Cosmic Leverage
Chavan's Cosmic Theory of Relativity (CTR)
Chavan's Theory of Light
Feeling Like a Physicist
TIME – The New Definition
$E \neq mc^2$: The Collapse of Classical Assumptions
The Chavan Codex

Echoes of War: The Russia-Ukraine Conflict and Its Global Ripples

Shattered Lives
The War Economy
Borders Redrawn
Arms and Allies
Collateral Damage

Justice in the Ashes

Somewhere in the Darkness of the Cosmos

The Blueprint of the Universe

The Cosmic Breath

From the Edge of the Universe

The Eternal Sage Chronicles

The Sage and the Sovereign

The Family Pillar

The People's Flame

The Warrior's Soul

The Bridging Mentor

The Guiding Light

The Exodus

The Last Cycle: 2075

The Fundamental Theory of Life & Existence

Energy in Action: The Soul Demystified

The Active Mind

Energy in Connection

Death: The Disconnection

Subconsciousness: The Chosen Disconnection

Desires: The Engine of Life

The Restoration of Human Knowledge

Restoring Bhagavad Gita

The War of Thoughts

Cracks in the Known

The Human Game

The Divine Dilemma

The Original Intent

The Answers

Ultra Modern Life Philosophy

The Duality of Laughter and Tears

Take It Easy

Compare to Accept

Standalone

It's Not AI, It's AHI - Amplified Human Intelligence

The Decision Paradox

The Four Sapiens

Malicious Script of Indian Polity

The IIT Legacy & Global Impact

The IIT Legacy & Global Impact

You are just a Version of Your Original

Win the Game You Didn't Choose

Always Keep Your Bags Packed

Reveal to Shield - Learn the Game of Facades

Engage Beyond Elections
Culture, Identity & Change: The Evolution of Indian Society
The Triangular Dynamics
Never Trust Without Doubt
If You Can't Change the Country, Leave the Country
Live Fearless with Fear
India: The Investment Magnet
Realignment
Powering a Billion Dreams
Universal Skill
Extension: From Brush to Algorithm
Human Side of Einstein
The Future Knows No Lines
Let's Bring the Generations to the Table
Unified Energy Dynamics
A Beat That Never Dies
Strategic Horizons
Right From Scratch
The Forgotten Innovators
Physics Fixing
The Savior Illusion
The Selfish Side
The Age of the Weakest Sapiens
Neural Wars
The Absorption: There Is No Death
Strange Therapist
Bubbles: Become a Drop
Beyond the Blue
Half-Scholar
The Law of Truth
Failure
Philosophy 2.0: A Structural Manifesto
Chaos: A Gentle Guide to the Unresolvable

Self-Help: New Syllabus
Rippling
Game: How to Play the Share Market
Out of Syllabus
Saffronized: The Code Behind the Crown
The Anger
Moksha: The Liberation
Chavan vs Einstein: The Universe Was Never Curved
Rocket Science: The Ripple Gambling
The Cosmos: The Still Universe
The Taste of Silence
Chavan's Post-Materialist Human Model
10-Second Rule
Chavan vs Hawking

Watch for more at https://www.sandeepjchavan.com/.